Modern Technologies in Ergonomics and Workplace Safety

The evolution of workplace technologies demands a balance between innovation and human-centric design. This book examines how organizations can create sustainable, efficient workspaces while prioritizing employee well-being and environmental responsibility. Through research and real-world applications, the book demonstrates how the thoughtful integration of technology with ergonomic principles leads to improved productivity and enhanced working conditions.

This book provides the tools and insights needed to create future-ready work environments. It presents insights into sustainable workspace design, incorporating Mediterranean materials and innovative approaches to virtual training. Readers will discover the transformative potential of wearable technologies, explore advanced visual analysis methods, and understand the implementation of crucial safety measures across various industries. The book provides practical frameworks for risk assessment, health monitoring, and construction safety, while highlighting the impact of Building Information Modeling (BIM) on workplace efficiency. By implementing the strategies outlined in these pages, readers can ensure that their organizations can create work environments that drive productivity while supporting both human and environmental well-being.

Modern Technologies in Ergonomics and Workplace Safety is written for researchers and graduate students seeking to expand their knowledge in workplace design and safety and pursuing innovative solutions in ergonomics. It will also appeal to policymakers responsible for shaping workplace regulations.

Human Factors in Design, Engineering, and Computing
Series Editors: Tareq Ahram and Waldemar Karwowski

This series focuses on research and development efforts intended to promote the comprehensive integration of people and technological systems. To help foster integration of human factors research with emerging technologies, this series will focus on novel methodologies, design tools and solutions that advance our understanding of the nature of human factors and ergonomics with intelligent technologies and services. It will address all aspects of human factors and human-centered design, with a particular emphasis on applications of emerging technologies, computing, Artificial Intelligence and Systems. The series offers a multidisciplinary platform for researchers and practitioners alike, discussing emerging issues in the field of human factors engineering systems, with a special focus on (but not limited to) computing and AI-based technologies. It will be useful for researchers, senior graduate students, graduate students, and professionals in different domains including human factors, engineering design, systems science and engineering, and ergonomics.

Human Factors and Emerging Technologies: Integrating People and Technological Systems
Edited by Tareq Ahram and Waldemar Karwowski

Human Factors in Intelligence, Technology and Analytics
Edited by Tareq Ahram and Waldemar Karwowski

Human-Technology: Blending Artificial Intelligence, Computing, and Intelligent Design
Edited by Tareq Ahram and Waldemar Karwowski

Converging Minds: The Creative Potential of Collaborative AI
Aleksandra Przegalinska and Tamilla Triantoro

3D Digital Design in Ergonomics and Human Factors
Tihomir Dovramadjiev

Artificial Intelligence and Human Existence
Edited by Akanksha Srivastava, A.V. Senthil Kumar, Disha Maheshwari and Samrat Ray

Modern Technologies in Ergonomics and Workplace Safety
Edited by Beata Mrugalska, Waldemar Karwowski and Tareq Z. Ahram

For more information about this series, please visit: www.routledge.com/Human-Factors-In-Design-Engnineering-and-Computing/book-series/%20CRCHFIDEC

Modern Technologies in Ergonomics and Workplace Safety

Edited by

Beata Mrugalska, Waldemar Karwowski and Tareq Z. Ahram

Routledge
Taylor & Francis Group
NEW YORK AND LONDON

Designed cover image: Shutterstock

First edition published 2026
by Routledge
2385 NW Executive Center Drive, Suite 320, Boca Raton FL 33431

and by Routledge
4 Park Square, Milton Park, Abingdon, Oxon, OX14 4RN

Routledge is an imprint of Taylor & Francis Group, LLC

First edition published by Routledge 2026

ISBN: 9781041136170 (hbk)
ISBN: 9781041138969 (pbk)
ISBN: 9781003672036 (ebk)

DOI: 10.1201/9781003672036

Typeset in Times
by Newgen Publishing UK

Contents

Preface

Advanced technologies reshape organizational structures along process lines and employees' need to balance various work processes. They positively impact jobs as long as they enhance productivity, streamline workflows and foster innovation. Their success depends on aligning technological advancements with ergonomics. A well-balanced approach that prioritizes both technological progress and human-centered design leads to a more sustainable, efficient, and engaging work environment. Moreover, sustainable development goals, when applied effectively, can lead to improved working conditions, reduced occupational risks, and greater efficiency in resource utilization. The integration of social and environmental agendas can benefit both employers and employees.

This book explores the intersection of ergonomics, sustainability, and workspace design, offering insights into how organizations can create workplaces that are not only efficient but also aligned with the broader objectives of sustainable development. It shows the role of virtual training in machinery operation and maintenance and the potential of wearable technologies for work system evaluation and ergonomic risk assessment. Additionally, it examines the application of visual analysis in complex static systems, qualitative expert evaluations through the GALA Framework, and workplace risks faced by machinery operators. Moreover, it is dedicated to health and safety, including discussions on muscle health as a predictor of COVID-19 mortality and the implementation of safety incentives in construction practices, particularly in developing countries. Furthermore, it highlights how technological advancements, such as Building Information Modeling (BIM), can bridge gaps in construction ergonomics and contribute to the global sustainability agenda. It presents cutting-edge research on the innovative use of Mediterranean materials for sustainable design.

This book is expected to be a valuable resource for graduate students, researchers, educators, teachers and policymakers interested in the fields of engineering and management. It aims to introduce innovative ideas that inspire readers to think critically while giving them the necessary knowledge and tools to formulate research questions for future studies.

About the Editors

Beata Mrugalska, Ph.D., D.Sc., Eur.Erg. is Associate Professor, Vice-dean for Science (2024–2028) at the Faculty of Management Engineering and head of the Division of Applied Ergonomics (2020–2024 and 2024–2028), Institute of Safety and Quality Engineering, Faculty of Management Engineering, Poznan University of Technology in Poland. She holds an M.Sc. in Management and Marketing from the Faculty of Mechanical Engineering and Management at the Poznan University of Technology (2001), and a Ph.D. in Machine Construction and Operation from the Faculty of Computer Science and Management at the Poznan University of Technology (2009). She was awarded D.Sc. (dr habil.) degree in Mechanical Engineering by the Faculty of Mechanical Engineering of the Poznan University of Technology, Poland (2019).

Waldemar Karwowski, Ph.D., D.Sc., P.E. is Pegasus Professor and Chairman, Department of Industrial Engineering and Management Systems and Executive Director, Institute for Advanced Systems Engineering, University of Central Florida, Orlando, USA. He holds an M.S. in Production Engineering and Management from the Technical University of Wroclaw, Poland (1978), and a Ph.D. in Industrial Engineering from Texas Tech University (1982). He was awarded a D.Sc. (dr habil.) degree in management science by the State Institute for Organization and Management in Industry, Poland (2004).

Tareq Z. Ahram, Ph.D. is Senior Executive and Research Professor in Emerging Technologies and Intelligent Healthcare Systems in the USA and a Visiting Professor in London, UK. He received a Ph.D. degree in Industrial and Systems Engineering from the University of Central Florida, USA, in 2008, with specialization in human systems integration and large-scale information retrieval systems optimization. M.Sc. in Industrial Systems Engineering – Human Factors from UCF in 2007, M.Sc. and B.Sc. in Industrial and Systems Engineering from the University of Jordan in 2002 and 2004, respectively.

Contributors

Alireza Abouhossein
Department of Ergonomics, School of Public Health and Safety, Shahid Beheshti University of Medical Sciences, Tehran, Iran

Clinton Aigbavboa
1cidb Centre of Excellence & Sustainable Human Settlement and Construction Research Centre, Department of Construction Management & Quantity Surveying, Faculty of Engineering and the Built Environment, University of Johannesburg, Johannesburg, South Africa

Amin Allafasghari, MD
School of Medicine, Ardabil University of Medical Sciences, Ardabil, Iran

Atabak Allafasghari, MD
School of Medicine, Ardabil University of Medical Sciences, Ardabil, Iran

Vesna Spasojevic Brkić
Faculty of Mechanical Engineering, University of Belgrade, Belgrade, Serbia

Ana Cardoso
ESAD Collage of Art and Design: Matosinhos, Porto, Portugal

Fatemeh Chaharaghran
Department of Industrial Engineering, Clemson University, Clemson, SC, USA

Fabian Dietrich
Westsächsische Hochschule Zwickau – University of Applied Sciences Zwickau, Zwickau, Germany

Daniel Dreyer
Airbus Defence and Space, Manching, Germany

Robert Eckardt
Hochschule Mittweida – University of Applied Sciences Mittweida, Faculty Engineering Sciences, Mittweida, Germany

Ana Teresa Gabriel
Faculdade de Ciencias e Tecnologia, Universidade Nova de Lisboa, Lisboa, Portugal

Leif Goldhahn
Hochschule Mittweida – University of Applied Sciences Mittweida, Faculty Engineering Sciences, Mittweida, Germany

Nemanja Janev
Faculty of Mechanical Engineering, University of Belgrade, Belgrade, Serbia

Stefanie Liebl
Westsächsische Hochschule Zwickau – University of Applied Sciences Zwickau, Zwickau, Germany

Marcus Löffler
Westsächsische Hochschule Zwickau – University of Applied Sciences Zwickau, Zwickau, Germany

Georgios Markopoulos
London School of Economics, Department of Management, London, United Kingdom

Panagiotis Markopoulos
Staffordshire University, London Digital Institute, London, United Kingdom

Torsten Merkel
Westsächsische Hochschule Zwickau – University of Applied Sciences Zwickau, Zwickau, Germany

Ivan Mihajlović
Faculty of Mechanical Engineering, University of Belgrade, Belgrade, Serbia

Mirjana Misita
Faculty of Mechanical Engineering, University of Belgrade, Belgrade, Serbia

Akash Nandi
Harvard University, Department of Economics, Cambridge, USA

Isabel L. Nunes
Faculdade de Ciências e Tecnologia, Universidade Nova de Lisboa, Lisboa, Portugal

Babatunde F. Ogunbayo
idb Centre of Excellence & Sustainable Human Settlement and Construction Research Centre, Department of Construction Management & Quantity Surveying, Faculty of Engineering and the Built Environment, University of Johannesburg, Johannesburg, South Africa

Kunle E. Ogundipe
1cidb Centre of Excellence & Sustainable Human Settlement and Construction Research Centre, Department of Construction Management & Quantity Surveying, Faculty of Engineering and the Built Environment, University of Johannesburg, Johannesburg, South Africa

Olatoyese Z. Oni
Department of Construction Management, Universiti Tunku Abdul Rahman, Kampar, Perak, Malaysia

Marie Oppitz
Airbus Defence and Space, Manching, Germany

Benedikt Petermeier
Airbus Defence and Space, Manching, Germany

Lorena Olmos Pineda
Universidad Nacional Autónoma de México, Mexico City, Mexico

Cláudia Quaresma
Faculdade de Ciencias e Tecnologia, Universidade Nova de Lisboa, Lisboa, Portugal

Sajjad Rostamzadeh
Department of Ergonomics, School of Public Health and Safety, Shahid Beheshti University of Medical Sciences, Tehran, Iran

Daniele Ruscio
Airbus Defence and Space, Manching, Germany

Raquel Salomé
Escola Superior de Artes e Design: Senhora da Hora, Porto, Portugal

Mascha Samkova
Sciences Po, Paris School of International Affairs, Paris, France

Catarina Santos
Faculdade de Ciências e Tecnologia, Universidade Nova de Lisboa, Lisboa, Portugal

Marvin Schopp
Airbus Defence and Space, Manching, Germany

Jorge Gil Tejeda
Universidad Autónoma Metropolitana Unidad Xochimilco, Mexico City, Mexico

1 Virtual Training for Operating and Maintenance of Machinery

Opportunities, Challenges and User Feedback

Robert Eckardt and Leif Goldhahn

1.1 INTRODUCTION

1.1.1 Background and Motivation for Virtual Learning

The digital transformation has brought about significant changes in almost all areas of life, which also affect professional qualifications in particular. The increasing flexibility of work tasks and the complexity of modern products pose both physical and mental challenges for employees.[1] Innovative technical and organizational tools are essential to meet these requirements.[2] Digital technologies, including virtual reality (VR), open up new potential for self-directed learning processes for employees in mechanical and plant engineering, one of the largest industrial sectors in Germany.

The safe operation and maintenance of machine tools and mechanical systems is one of the skills critical to success in this sector. Currently, VR-based simulators for these applications only exist to a very limited extent.[3] This is particularly worrying as the potential for damage in the event of accidents or operating errors can be enormous. A clear example of the successful use of VR technologies can be found in civilian pilot training, where training simulators make it possible to practice realistic scenarios without risk. These simulators not only promote the learning of specific operating maneuvers but also enable an analysis of the behavior of the trainees, whereby potential misconduct can be identified and preventively remedied.[4]

As part of the Living Lab "Hybride Arbeitssysteme" at University of Applied Sciences Mittweida, a virtual, learning-promoting training course on the use and maintenance of machine systems is being designed using the example of the "Trainingsfabrik 4.0" flexible manufacturing system.[5] This training follows an integrative approach that takes into account both technical and mental aspects and is based on current scientific findings and practical requirements. The aim is to create an effective learning environment that enables participants to continuously improve

DOI: 10.1201/9781003672036-1

their skills. Through interactive simulations, various scenarios can be played out – from routine maintenance processes to emergency situations –promoting an in-depth understanding of the functioning, operation and maintenance of the systems.

The targeted use of modern technologies and a methodically sound development approach are crucial to ensuring that these virtual learning scenarios meet the needs of users and offer real added value.[6]

1.1.2 Relevance of the Topic in Digital Change

The digital transformation has affected all areas of society, and industry is also facing enormous changes. The integration of digital technologies into manufacturing and maintenance processes is no longer optional but mandatory in order to remain competitive.[7] Virtual learning environments can be a key to this transformation, as they not only offer new opportunities for innovative learning but also help to train skilled workers effectively.

The relevance of this topic is also reflected in current developments on the labor market: companies are increasingly looking for employees who have digital skills and an in-depth understanding of modern technologies. Virtual learning can help to promote these skills in a targeted manner and thus enable employees to work flexibly and safely, which also contributes to the company's resilience.

1.2 BASICS

1.2.1 Digital Transformation in the Industry

Digital transformation describes the process of integrating digital technologies into all areas of a company.[8] In industry, this includes not only the automation of production processes but also the introduction of new business models and innovative approaches to employee training. This transformation is being driven by various factors: increasing competitive pressure, changing customer requirements and technological advances such as the Internet of Things (IoT), big data and artificial intelligence (AI).

A key requirement of this transformation is continuous, more individualized employee training.[9] Traditional seminar-based training methods often reach their limits here. They are time-consuming and rarely offer the opportunity to apply what has been learned in practice. Virtual learning environments, on the other hand, enable employees to access training content virtually at any time and from any location and to use it interactively.

1.2.2 Virtual Reality (VR) Definition

VR has established itself as a technological tool that brings the way in which information is perceived and processed closer to people's natural sensory perception. In the literature, VR is defined as the "visual presentation and manipulation of 3D data in real time."[10] This definition implies not only a technical dimension but also a deeper interaction between the user and the virtual world presented. Goldhahn[11] identifies three central properties that characterize VR:

Immersion – this refers to the realistic presentation and perception of content, which enables the user to immerse themselves in an "other" reality.
Interaction – this refers to the user's ability to influence the virtual environment in real time and actively change it.
Imagination – this characteristic describes the experience of a quasi-real world in which users can develop their imagination.

The use of VR appeals to various human senses: visual, acoustic and/or haptic perceptions are stimulated. The precise visualization and the possibility of perspective changes help users to develop a realistic perception of the virtual environment.[12]

VR applications offer numerous advantages in the field of education and training: they enable realistic simulations of complex processes without risk to humans or machines. VR can help to identify potential hazards at an early stage and develop appropriate strategies for action, particularly in safety-related topics such as the maintenance of machine tools.

In recent years, studies have shown that the VR systems available on the market, such as head-mounted displays (HMD), differ in terms of the degree of separation of the user from the physical real world and their integration into a virtual environment.[13] HMDs such as the HTC Vive Pro offer a high degree of immersion.[14] They allow users to interact with both hands via controllers in a virtual learning environment. This allows work tasks, including specific operating actions, to be trained in a targeted manner (see Section 1.5.4).

1.2.3 Didactic Aspects in the Context of Virtual Training

The didactic-oriented design of virtual learning scenarios is of central importance for the success of employee training. In an increasingly digitalized world of work, effective learning methods are essential to promote employees' skills and prepare them for real-world challenges.

To better understand the different learning methods and their specific characteristics, it is helpful to present them in a table. Table 1.1 shows how different learning methods can be classified according to their characteristics in terms of machine-relatedness, process-relatedness, product-relatedness, data orientation and other dimensions. This overview makes it possible to select the appropriate methods and recognize their strengths and weaknesses in context.

The integration of different learning approaches makes it possible to design comprehensive and practical training. A particularly effective approach is self-directed, self-organized and virtually self-experienced learning, in which users can explore virtual environments independently. This method promotes not only personal responsibility but also the participants' problem-solving skills. In addition, situational learning offers the opportunity to design scenarios that simulate real work situations.

TABLE 1.1
Application Characteristics of Learning Methods for Industrial Activities

Learning Method	Machine-Related	Process-Related	Product-Related	Data-Oriented	Visual	Participating	Cooperative	Self-employed	Flexible in Location	Time Flexible
Lecture, instruction	o	o	o	+	-	+	-	-	-	-
Educational talk	o	o	o	o	-	-	+	-	-	-
Computer or web-based learning	o	o	-	o	o	+	-	+	+	+
Blended learning	o	o	-	o	o	+	o	o	o	o
Multimedia Process Descriptions	+	+	+	+	+	-	-	+	+	+
Four-step method (prepare, demonstrate, perform, practice)	o	o	o	o	o	o	+	-	-	-
Guiding text method (informing, planning, deciding to execute, executing, self-monitoring, self-assessment)	o	o	o	o	o	-	o	+	-	o
Virtual Training	+	+	+	o	+	-	o	+	+	+
Job enlargement, job enrichment, job rotation	+	+	+	o	o	o	o	+	-	-
Tutorials	o	o	o	o	o	o	-	-	-	-
Assistance, substitution	+	+	+	o	o	-	+	+	-	-
Small group work	o	o	o	o	o	-	+	-	-	-
Role play, case study, business game	o	o	o	o	o	-	+	-	o	o
Training Area, Training Plant	+	+	+	+	+	-	+	o	-	-
Exchange of experience	o	o	o	o	-	-	+	-	-	-
Problem-solving in groups	o	o	o	o	-	-	+	-	-	-

Key + applies o partially applicable - not applicable

Note: Copyright: Hochschule Mittweida.

This creates an authentic learning experience that allows users to test and develop their skills in a safe environment.

Numerous proven learning methods for industrial activities are available in the run-up to virtual training or in preparation for work. These include lectures and instructions, computer-based training (CBT) and web-based training (WBT). Blended learning combines face-to-face and online formats, allowing the learning process to be designed flexibly. Multimedia process descriptions and instructional discussions provide additional support in understanding complex processes. Multimedia descriptions describe activities and provide context-related, process-relevant data, for example, setting parameters.

The main disadvantage of all these learning applications is that practical knowledge and skills can often only be taught using real objects. The fundamental added value of VR-based technologies "lies in the temporal and spatial flexibility of the learning processes."[15]

Overall, it can be seen that virtual training formats offer a wide range of opportunities to specifically promote employees' skills in plant operation and maintenance. At the same time, however, they also present challenges, whether in terms of technical requirements or the acceptance of new learning methods by users. User feedback is therefore of great importance for the continuous improvement of these training formats.

The following section outlines the specific opportunities and challenges of virtual training. Technical aspects in terms of work tasks on machine systems and user feedback are considered.

1.2.3 Virtual Training

Virtual training is based on teaching learning content using virtual techniques. The provision of relevant action situations enables employees to gain practical experience and develop their skills in a safe environment.[16]

The advantages of virtual training are manifold. The specialist literature shows that it has not only a high level of practical relevance but also significant improvements in learning effectiveness.[17,18] The use of VR technologies promotes the teaching of process-relevant skills and creates realistic training environments. These environments not only support orientation in real, physical space but also increase learning motivation and acceptance among users.[19,20] The machines and systems can be procured at a later date or used in parallel to create value.

The design of virtual learning scenarios as a didactic component of virtual training is based, for example, on the operating states of machine tools. These include normal operating conditions as well as specific situations such as start-up and shutdown under shift-related conditions and commissioning and decommissioning. Operational activities such as machine set-up are also taken into account in these scenarios.[21,22]

An illustrative example of the benefits of such training is the virtual scenario for maintaining a machine. Here, learners can simulate complex processes and identify and minimize potential sources of error. This not only leads to a reduction in the error rate during real maintenance work but also contributes to the safety of employees and technical equipment.

In previous work, three different types of virtual learning scenarios were developed that are specifically geared toward training for commissioning a milling machine within a flexible manufacturing system. These scenarios vary in terms of their learning requirements and were implemented using the HTC Vive Pro VR headset. The qualification was carried out both on the hardware side by the VR headset and on the software side by an in-house development of the Mittweida University of Applied Sciences.

In the context of machine tool maintenance, this virtual training plays a decisive role in the efficiency and longevity of industrial production processes. Maintenance is an essential part of a company's operational management and has a direct impact on efficiency, costs, safety and product quality. Maintenance must be viewed as a strategic process whose objectives include minimizing technical faults and downtime and maximizing equipment availability and service life.[23] Virtual training in this area of maintenance contributes to increasing product quality through self-managed maintenance activities, reduces the reject rate, promotes environmental protection and increases occupational safety. It also supports the reduction of inventories and contributes to scheduled production and personnel deployment.

1.3 STATE OF THE ART

1.3.1 Technical Aids to Support Processes for the Commissioning, Use and Maintenance of Mechanical Systems

Technical aids play a central role in supporting manufacturing processes in industrial environments. Basic tasks in manufacturing are shown in Figure 1.1.

Traditional classroom training methods usually lack the system specifics. In industrial operations, often only operating instructions and maintenance instructions

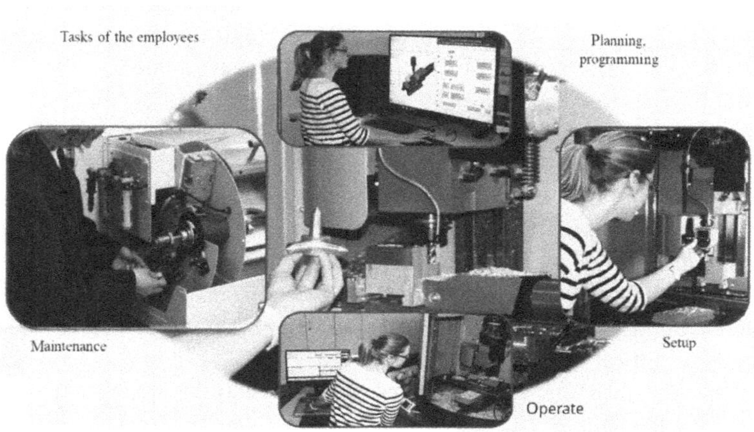

FIGURE 1.1 Basic tasks of employees in production. Copyright: Hochschule Mittweida.

from the manufacturers of individual components are available. Digital training materials can create a link between manufacturer knowledge and operational experience and specifications on the one hand and the ready-to-use knowledge of production employees on the other. Both hardware and software solutions are crucial for increasing efficiency, minimizing downtimes and eliminating hazards.

Important technologies such as sensors and specialized maintenance software are explained using the example of maintenance tasks. Sensors continuously record data on the condition of machines, enabling precise monitoring of performance. Localizable IoT sensors, which are integrated directly into machines, collect real-time data that is essential for the analysis and optimization of maintenance strategies. Predictive maintenance tools use these data analyses to predict potential failures and initiate proactive maintenance measures.

Mobile apps provide technicians on site with access to important information and support them in carrying out maintenance work. Augmented reality technologies can support technicians in troubleshooting and repairs by overlaying visual instructions directly at the workplace.

The use of VR, on the other hand, does not yet play a significant role in maintenance, although the advantages, such as training the installation and removal of components on machine tools, bring many benefits. For example, it is possible to train maintenance personnel on what to look out for before the actual maintenance work is carried out, without taking the real machine out of operation. On the one hand, this has the advantage that maintenance and repair work can be carried out more easily, saving valuable production time. On the other hand, this extends the machine running time, as the real machine does not have to be shut down for training purposes.

VR should therefore be integrated into training programs in order to continuously improve the knowledge and skills of employees in the production area. Virtual learning environments allow employees to gain hands-on experience in a safe environment. These training methods provide not only a cost-effective solution for improving employees' skills but also the opportunity to better overcome challenges when dealing with complex machines.

1.3.2 Virtual Reality Technology: Overview of the Hardware and Software Used

University of Applied Sciences Mittweida has been working intensively on VR technologies and has tested various hardware and developed software solutions (Figure 1.2):

- Hardware:
  A VR Powerwall enables groups of planners, machine users and maintenance staff to jointly design, discuss and test ideas and training programs semi-immersively.
  VR glasses such as HTC Vive (see Figure 1.2) and HTC Vive Pro offer fully immersive experiences for individual learners and, if required, a conditional view for trainers and viewers.

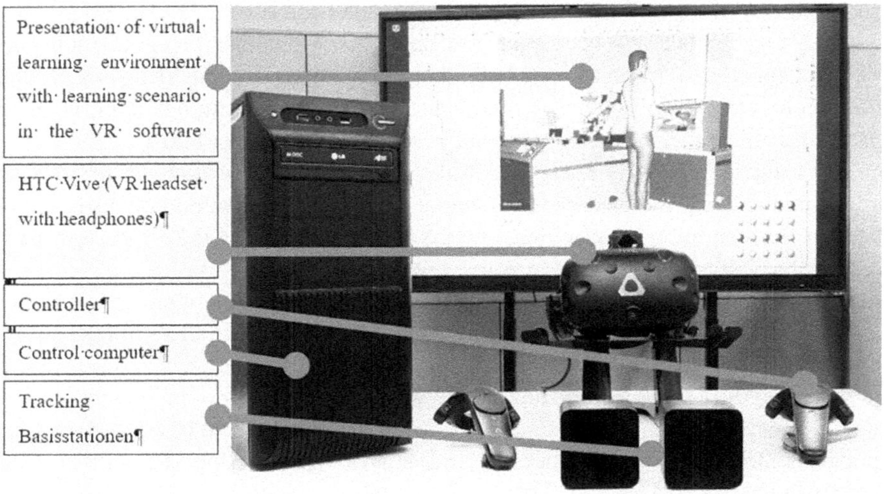

FIGURE 1.2 Hardware (HMD) and software used for virtual training. Copyright: Hochschule Mittweida.

- Software:
  Specialized programs such as IC.IDO for creating virtual learning scenarios make it possible to adapt content to specific systems and individual learning needs.

These technologies form the basis for innovative qualification for the manufacturing sector.

1.4 IMPORTANCE OF INTERACTION AND CHALLENGES WHEN USING TECHNICAL DEVICES IN THE LEARNING ENVIRONMENT

The interaction between people and technical devices plays a crucial role in the success of virtual learning. An intuitive user interface and clear feedback during training are essential for users to feel comfortable and learn effectively. A positive user experience (UX) motivates users to engage more intensively with the learning material, which has a direct impact on learning success. Various measurement methods (e.g., User Experience Questionnaire, A/B tests, System Usability Scale) are used in the specialist literature to quantify the success of UX.[24]

Firstly, user-friendly interfaces such as real-time stereoscopic graphic output enable users to absorb information well and quickly to achieve their goals. Smooth interactions promote user well-being and increase the likelihood of continued use of the learning application. Secondly, software should be designed to be accessible to all users, including people with disabilities such as color blindness. A good UX takes into account different needs and abilities. Third, good interactions provide immediate feedback on inputs; this helps users understand if their actions were successful or if adjustments are needed.

Virtual training environments can achieve these positive effects. However, challenges can be found in the operation of virtual machines within virtual environments. Technical barriers are an obstacle, as not all users have the necessary VR skills; therefore, targeted training must also be offered for this.

The InnArbeit Institute at University of Applied Sciences Mittweida developed its own standardized training scenario for the subsequent use of HMDs and controllers during virtual training. This is placed before the actual training and achieved very good results with inexperienced VR users.[25] Physical limitations, for example, in the area of the eyes, can lead to some users having difficulty performing certain movements (motion sickness). These challenges must be taken into account during development in order to achieve broad user acceptance.

1.5 DEVELOPMENT OF VIRTUAL LEARNING SCENARIOS

1.5.1 Methodology of Learning Scenario Development

The development of an effective learning scenario requires a structured and systematic approach that can be divided into several phases. Figure 1.3 shows the methodical approach to the systematic development of virtual learning scenarios.

First of all, the needs assessment is of central importance, as it identifies the specific requirements for the training. This is followed by conceptualization, in which the scenario is designed taking didactic principles into account. The subsequent implementation comprises the technical implementation using suitable software tools, followed by a thorough evaluation, during which a test phase is carried out to check the learning scenario against defined criteria. This iterative process ensures continuous improvement of the training offer based on user feedback.

A key advantage of virtual learning environments is the greatest possible transparency and support for all participants. The targeted use of a small number of specific tools not only develops basic media skills but also promotes learner motivation. In addition, reflection on an individual and collective level enables a deeper examination of one's own learning progress.

Nevertheless, there are also some disadvantages associated with the use of virtual learning scenarios. These include social isolation, distraction and excessive demands as well as access barriers. The quality and authenticity of the content can also be critically questioned.

In the context of virtual learning scenarios, numerous action situations are provided in the sense of problem-solving or action strategies.

The literature[26,27] emphasizes that the use of VR in learning scenarios in particular leads to a high level of practical relevance, and good results can be achieved in terms of learning effectiveness.

Virtual learning scenarios not only support the teaching of procedural and action-related skills but also offer realistic environments that serve both to provide orientation in real space and to increase learning motivation and acceptance.

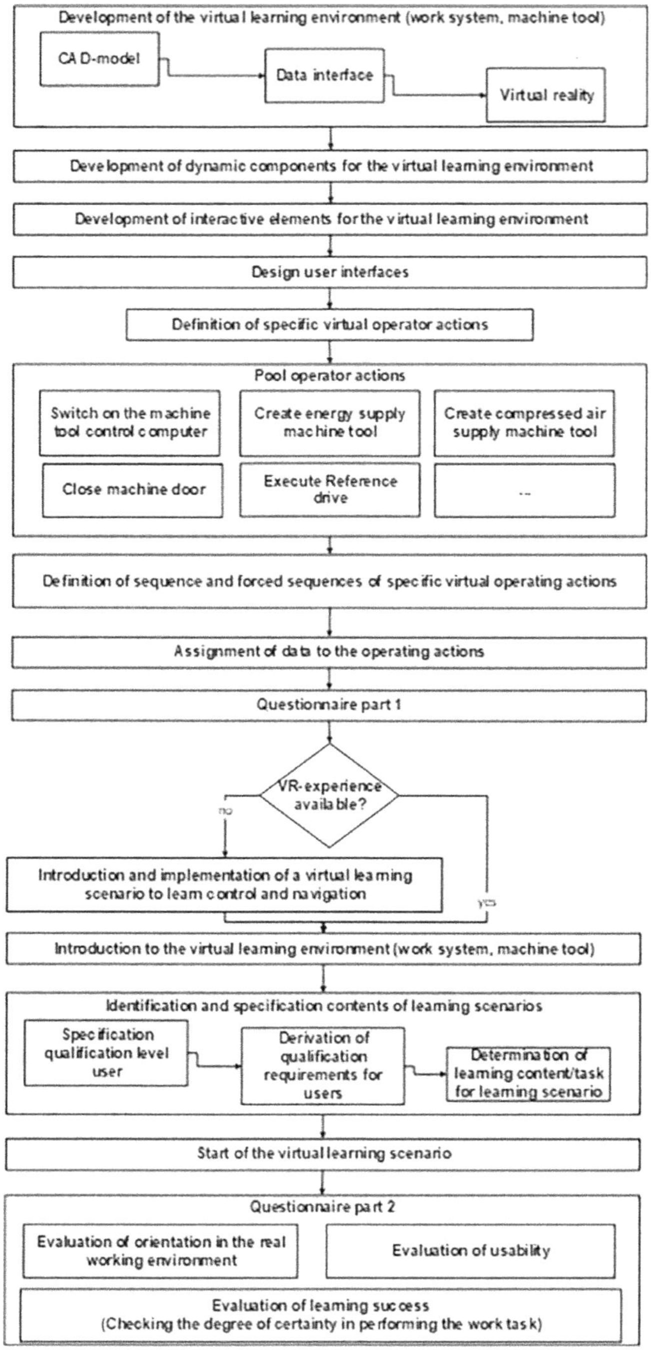

FIGURE 1.3 Methodological approach to the development of a virtual learning scenario. (Goldhahn, Leif; Eckardt, Robert: Learning Scenarios for the Improvement of Operating Safety of Machine Tools. In: Black, N.L.; Neumann, W.P.; Noy, I. (eds.): Proceedings of the 21st Congress of the International Ergonomics Association (IEA 2021). IEA 2021. Lecture Notes in Networks and Systems, vol. 223. Cham: Springer, pp. 749–760, 2021. https://doi.org/10.1007/978-3-030-74614-8_92.)

These virtual scenarios are based on the operating states of machine tools, such as normal operation or start-up and shutdown, as well as on the activities required for part-specific machining, such as set-up or operational activities. Work tasks such as the commissioning of a machine can be mapped and trained within such a virtual learning scenario. This helps to reduce the number and severity of errors during real commissioning, which protects both man and machine from danger.

1.5.2 Operating Level and Operating Complexity of Virtual Learning Scenarios

In preliminary work, three types of virtual learning scenarios were developed and categorized into three times three levels. They take into account different learning requirements for virtual training. The qualification is carried out on the hardware side using VR glasses (HTC Vive Pro) and on the software side using an in-house development by the InnArbeit Institute at Mittweida University of Applied Sciences. The procedure for creating such virtual learning scenarios (see Section 1.5.1) makes it possible to develop virtual training courses based on dynamic 3D models.[28] These training courses train users in the operation and execution of activities on machine tools.

The methodology of Goldhahn et al.[29] proves to be particularly helpful in determining the required qualification for virtual learning scenarios. This methodology uses a step-by-step model to provide suitable learning content for different levels of required and existing employee skills. According to Goldhahn et al.,[30] both the operating complexity (OC) and the operating level (OL) were defined in three levels in order to structure and classify the complexity of virtual training.

Based on this methodology, a virtual training course was first created for commissioning a milling machine and the associated operating actions. The virtual learning scenario offers information-guided operating actions, supported by visual, auditory and haptic elements. In this training, the user navigates independently through the learning environment and performs the respective operating actions in a predefined sequence; errors in this sequence are excluded, which ensures the successful completion of the training.

Table 1.2 visualizes the stage model for determining the qualifications required for virtual learning scenarios according to Goldhahn et al.[31]

The type and number of activities (information) required for the successful implementation of the learning scenario, which are performed by the user (input) is determined by the OL. Furthermore, the OC includes the degrees of freedom and dependencies in the learning environment available for the respective virtual training.

Based on the approach described in Section 1.5.1[32] and the level model developed, three types of learning scenarios with different learning requirements can be defined. These were designed for the required operating actions when switching on a milling machine in the Training Factory 4.0 learning environment:

- Learning scenario 1: Completely guided operating actions (visual observation, without own actions (OC1/OL1))

TABLE 1.2
Step-by-Step Mode: l of VR-Based Operating Actions for Classifying Learning Scenarios

OL = operator level	Level	3	OC OL	1 3	OC OL	2 3	OC OL	3 3
Improve the level of information provided for the specific operator action		2	OC OL	1 2	OC OL	2 2	OC OL	3 2
		1	OC OL	1 1	OC OL	2 1	OC OL	3 2
			1		2		3	
			OC operator complexity Improve the level of control and the number of decisions made by the user					

Source: Goldhahn, Leif; Eckardt, Robert: Learning Scenarios for the Improvement of Operating Safety of Machine Tools. In: Black, N.L.; Neumann, W.P.; Noy, I. (eds.): Proceedings of the 21st Congress of the International Ergonomics Association (IEA 2021). IEA 2021. Lecture Notes in Networks and Systems, vol. 223. Cham: Springer, pp. 749–760, 2021. https://doi.org/10.1007/978-3-030-74614-8_92.

- Learning scenario 2: Information-guided operating actions (visual and audio-supported operating actions (OC2/OL2))
- Learning scenario 3: Completely autonomous operating actions (error possibilities and feedback for users (OC3/OL3))

The virtual learning scenarios are described in the following sections.

1.5.3 Virtual Learning Scenario for Commissioning a Machine Tool

An exemplary virtual learning scenario includes the commissioning of the milling machine, which is located in the Training Factory 4.0 at Mittweida University of Applied Sciences. This virtual learning scenario is based on the step-by-step model described in Section 1.5.2 and contains information-guided operating actions that include visual and audio-supported operating actions (OV2/OL2). Figure 1.4 shows the test person carrying out the virtual learning scenario. Within the virtual learning environment, she is shown as an avatar and has the task of following the footprints on the floor. These are used for orientation and navigation to the next operation to be carried out. In addition, the illustration shows information panels that are made available to the test person and contain instructions on the action to be performed (What to do? Why? How?).

In this scenario, the user navigates through the learning environment themselves and works through the respective actions virtually in the predefined sequence. Errors in the sequence of actions are not possible, which ensures the successful completion of the scenario.

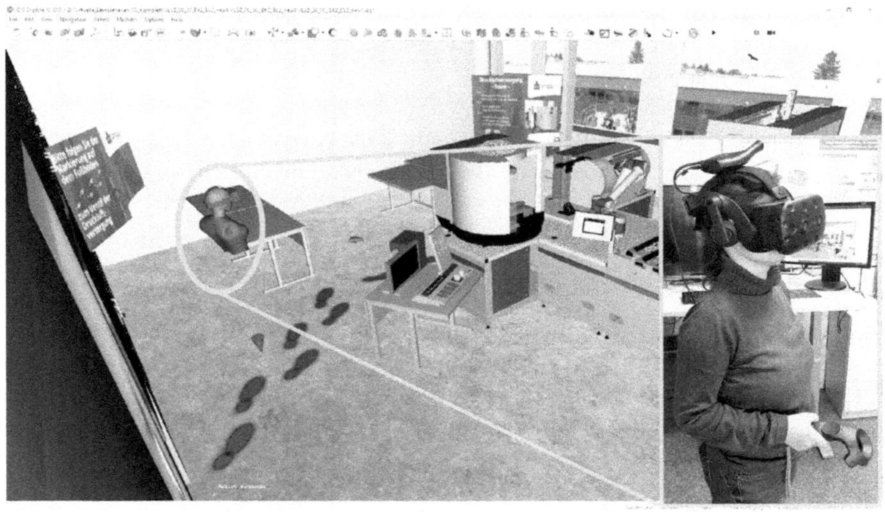

FIGURE 1.4 VR-based learning scenario – information-guided operating actions. (Goldhahn, Leif; Eckard, Robert; Müller-Eppendorfer, Katharina: Standardisiertes Übungsszenario für HMD-basierte virtuelle Lernumgebungen. In: Ökologische Transformation in Technik, Wirtschaft und Gesellschaft? Scientific Reports. 26. Interdisziplinäre Wissenschaftliche Konferenz Mittweida, ISSN 1437-7624, Nr. 2, S. 105–108, 2021. https://doi.org/10.48446/opus-12300.)

The advantages and disadvantages of the learning scenario milling machine commissioning (OC2/OL2) are shown below:

Advantages:

(+) System-side process-oriented guidance of learners
(+) Users can familiarize themselves with their learning environment (here Trainingsfabrik 4.0)
(+) Less effort is required to explain the operating action to be performed
(+) Little effort is required to explain the spatial position of the next operating action
(-) Occasional assistance is required for users when learning how the controllers work for navigation

Disadvantages:

(-) Occasional assistance for users when learning the virtually mapped operating actions (e.g., turning switches)
(-) Learning from mistakes in operating sequences is not part of learning scenario two

(-) Learning how the controllers work is necessary for navigation (may be problematic for achieving the learning objective because frustration arises from incorrect navigation)

The disadvantages identified by the users after carrying out the learning scenario made it necessary to modularize the operations to be carried out in the learning scenario.

1.5.4 Modularization of Operating Actions

The term "virtual reality" was defined in Section 1.2.2. In addition, VDI 3699 defines the term "operation" as a "process in which an operator influences the operation of an already selected technical device [(e.g. machine)] or stimulates the execution of a function."[33] According to Hacker,[34] an action is understood as "a self-contained unit of activity."

In summary, VR-based operating actions refer to a comprehensive virtual process for initiating, performing and completing a function on a virtual technical device, whereby these actions can have a direct and multi-stage effect on the virtual technology.

VR software and fully immersive VR glasses were used to represent a quasi-real world in the form of a learning environment that can be influenced in real time by virtual operating actions. Table 1.3 provides an example of a hierarchical insight into a VR-based operating procedure that was implemented within the virtual learning environment *Trainingsfabrik 4.0* supply for a machine tool.[35]

This table also shows the various elements (task, activity, action, operation and movement) of a virtual user action. This hierarchical classification enables a methodical structuring of the virtual user actions.[36] In this way, a systematic and modular provision of VR content is realized to cover the qualification needs of employees.[37]

An HMD, such as the HTC Vive Pro, with a high degree of immersion enables free interaction with both hands using controllers at a virtual workplace, allowing work tasks including operating actions to be trained in a targeted and repeated manner.

1.5.5 Transfer and Validation for Assembly Tasks and VR PowerWall

Based on the modularization of operating actions, a virtual learning environment for assembly tasks was developed in order to test the benefits of modularization for more complex operating actions.[38]

The methods and techniques for virtual training were further validated. Encouraged by this positive application, new software-side functions were tested and initially modeled in a virtual learning environment for the PowerWall. Significant differences in good design were also evident here, resulting from the different interaction and the one-sided (Powerwall) versus all-sided projection (HMD). This played almost no role in the individual workstation for the assembly of small products. For virtual movement and the perception of descriptions or signal lights during a

TABLE 1.3
Modularization of VR-Based Operating Actions Using the Example of Producing the Energy

Operation	Description	Information
Task	Establish a power supply for the switched-off milling machine	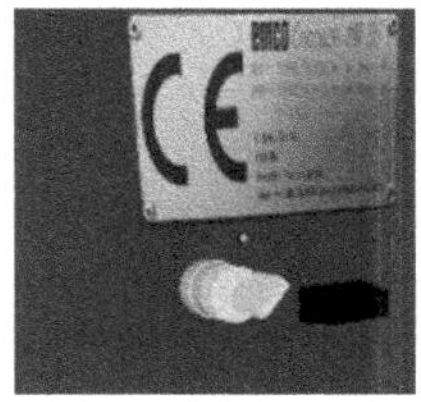
Activity	Turning on the milling machine	
Action	Turning the rotary switch	
Procedure	Manual rotation of the rotary switch clockwise until the end stop	
Movement	Reach for the rotary switch and operate it	

maintenance task, however, this becomes significant. There are several necessary viewing directions, and the required information should be clearly visible in each case. They must therefore be decentralized and deliberately aligned and positioned in the room. VR-based operating actions offer significant advantages, such as the virtual testing of a machine tool instead of abstract frontal instruction or the consideration of different levels of knowledge among employees. In the further course of the development work, it must be determined how pronounced the aforementioned advantages are. To this end, questionnaires are to be issued and evaluated after the VR-based operating procedures have been used.

1.5.6 Virtual Learning Scenario for the Maintenance of a CNC Lathe

Figure 1.5 illustrates virtual training using the example of maintaining a CNC lathe within a newly developed learning environment. Based on the categorization of virtual learning scenarios in Section 1.5.2, this scenario is in the "CL2/OL2" category. The locations where the respective actions are carried out are identified by markings such as boxes around screws. In the case of complex processes, animations illustrate the action process accordingly. The process is therefore demonstrated virtually and then carried out virtually by the learner. If the virtual execution is successful, a progress bar above the information board indicates positive feedback. The next step is then displayed on the board.

Findings from tests show both advantages and disadvantages of virtual training. The advantages include system-based, process-oriented guidance using information boards and less effort when explaining complex operating actions in real environments. The creation of a stable virtual learning scenario, on the other hand, requires a relatively high level of effort.

The tests were carried out with 30 people who each completed the virtual training and were then given a questionnaire.

The questionnaire included the following questions and criteria, which the test subjects were asked to assess:

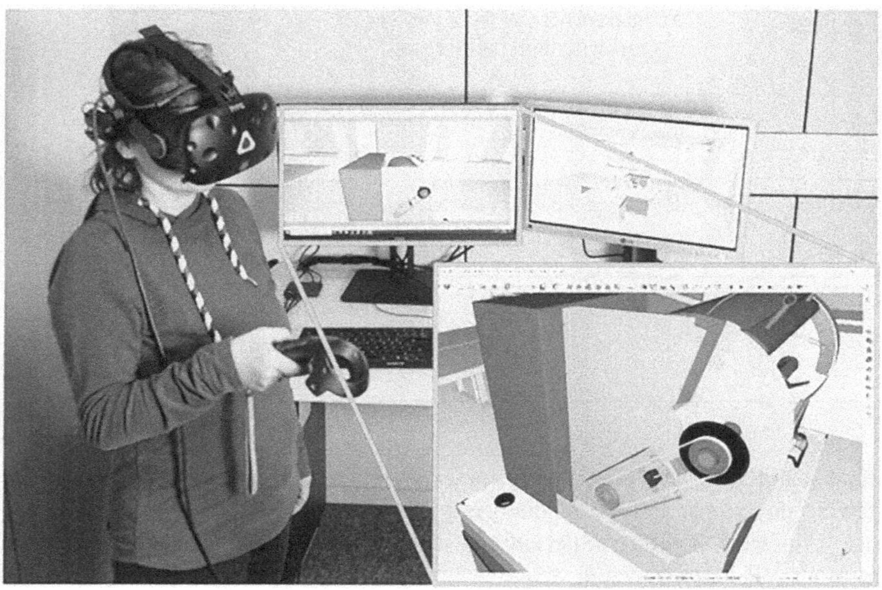

FIGURE 1.5 VR-based learning scenario – maintenance of a lathe. (Eckardt, Robert; Goldhahn, Leif: Virtual Training for the Maintenance of Machine Tools. In: Isabel L. Nunes (eds.) Human Factors and Systems Interaction. AHFE (2024) International Conference. AHFE Open Access, vol. 154. AHFE International, USA, pp. 199–207, 2024. http://doi.org/10.54941/ahfe1005367.)

- The virtual learning scenario has helped you to develop a clear spatial image of the real training factory
- The virtual learning scenario improved your ability to correctly understand and repeat the operating procedures in the real plant, such as connecting the power supply
- The information boards in the virtual learning scenario were easy to read

The test subjects were asked a total of 15 questions. Basically, the criteria could be assessed in five levels (strongly disagree, disagree, partly/partly agree, agree, fully agree). In addition, questions were asked that dealt with verbal responses regarding the potential use of virtual training or an alternative use.

The aim of the questionnaire was to obtain information about the user experience in order to significantly increase usability for other users.

Table 1.3 and Figure 1.5 contain statistical values and the resulting distribution for the question "The virtual learning scenario helped you to develop a clear spatial image of the real training factory" and show that the majority of test subjects had a very positive attitude toward the question. With a mean value of 4.57 and a median of 5, it can be seen that most test subjects agreed or fully agreed with the statement. The low value of the standard deviation shows that there is little variance in the answers, which means that the opinions of the respondents are largely in agreement. It can be concluded from this that the results indicate that the test subjects have a positive

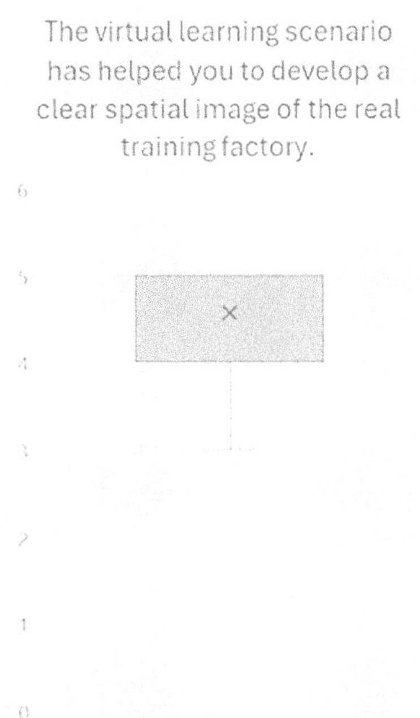

FIGURE 1.6 Evaluation of the survey on the transferability of the virtual experience to the real world.

TABLE 1.4
Overview of Static Values for the Question of Spatial Perception

Statistics

Question 10

					Frequency	Percent	Valid Percent	Cumulative Percent
N	Valid	28						
	Missing	2						
Mean		4.5714						
Median		5.0000	Valid	Partly/partly	1	3.3	3.6	3.6
Std. Deviation		0.57275		Agree	10	33.3	35.7	39.3
Minimum		3.00		Fully agree	17	56.7	60.7	100.0
Maximum		5.00		Total	28	93.3	100.7	
Percentiles	25	4.0000						
	50	5.0000	Missing	System	2	6.7		
	75	5.0000	Total		30	100.0		

opinion of the question posed overall. The fact that no one gave a rating of 1 or 2 underlines the strong approval of the group.

1.6 SUMMARY AND OUTLOOK

The development of virtual learning scenarios requires a structured, multi-stage approach that begins with needs assessment and extends through conceptualization, implementation and evaluation. An iterative process enables continuous improvement based on user experience (UX). Virtual learning environments offer advantages such as transparency, support and the promotion of media skills and learner motivation. Nevertheless, there are challenges such as access barriers in the sense of operating problems due to the hardware-side input devices.

The methodology for creating virtual learning scenarios involves categorizing them into three levels in order to take different learning requirements into account. One example is the virtual training for commissioning a milling machine, which is supported by visual and auditory elements. The modularization of operating actions enables a structured approach to complex tasks and promotes the qualification of employees.

Another example is the virtual training for the maintenance of a CNC lathe, which is supported by animations and progress displays. The tests show both advantages and disadvantages of virtual training, although the effort required to create stable scenarios is considered challenging.

A three-stage teaching/learning concept has been developed as a result of the experience gained from virtual training for lathe maintenance. This begins with a multimedia process description, which primarily contains dates and backgrounds, but also images and animations selected from the virtual learning environment. This leads to repetition and recognition effects for the learner. The virtual training can possibly lead to a test. If this is successfully completed, training on the real machine follows. This involves size and space conditions, accessibility, visibility and manual tasks that are physically more feasible. This is the most time-consuming part of learning and should be omitted if possible. A comparison group and measurable criteria can be used for evaluation.

In the further course of the development tasks, a method is to be developed that helps to clarify the question of why errors are made with regard to hardware inputs (e.g., incorrect pressing of buttons on the controllers) but also software-side errors (e.g., incorrect pressing or rotation of virtual objects). With the help of this method, a tool is to be developed that can document and analyze incorrect user inputs on the software side. Based on this, a qualitative statement should be made about how a virtual learning environment can be designed in a systematic, user-friendly, and, above all, learning-promoting way.

ACKNOWLEDGMENT

The authors would like to thank the Federal Ministry of Education and Research for its financial support and the Project Management Agency Karlsruhe (PTKA) for its support of the research project "PerspektiveArbeit Lausitz (PAL) – Kompetenzzentrum für die Arbeit der Zukunft in Sachsen und Brandenburg," funding code 02L19C300.

NOTES

1 DIN EN ISO 6385, Grundsätze der Ergonomie für die Gestaltung von Arbeitssystemen (2016).
2 Bundesministerium für Arbeit und Soziales; Digitalisierung am Arbeitsplatz, S. 16, Zugang 08.01.2020, 2016. www.bmas.de/SharedDocs/Downloads/DE/Publikationen/Forschungsberichte/fb-468-digitalisierung-am-arbeitsplatz.pdf?__blob=publicationFile&v=2.
3 Hirt, C.; Spahni, M.; Kompis, Y.; Jetter, D.; Kunz, A.: Alles rund um die Maschine – Begehbare virtuelle Schulung an Werkzeugmaschinen. In: VAR² 2019 – Realität erweitern, Hrsg. Putz, M.; Klimant, P.; Klimat, F.; Chemnitz, ISBN: 978-3-00-064420-7, S. 43–58, 2019.
4 FOERST GmbH, Effiziente Fahrsimulatoren für Pkw, Lkw und Busse. Zugang am 07.01.2020. www.fahrsimulatoren.eu/de/.
5 Goldhahn, Leif: Entwicklung hybrider Arbeitssysteme. In: Entwicklung hybrider Arbeitssysteme. Scientific Reports. Hochschule Mittweida, ISSN 1437-7624, 2023, Nr. 1, S. 1–5, 2023. https://doi.org/10.48446/opus-14344.
6 Goldhahn, Leif; Eckardt, Robert: Learning Scenarios for the Improvement of Operating Safety of Machine Tools. In: Black, N.L.; Neumann, W.P.; Noy, I. (eds.): Proceedings of the 21st Congress of the International Ergonomics Association (IEA 2021). IEA 2021. Lecture Notes in Networks and Systems, vol. 223. Cham: Springer, pp. 749–760, 2021. https://doi.org/10.1007/978-3-030-74614-8_92.
7 DIN EN 13306: Instandhaltung – Begriffe der Instandhaltung; Dreisprachige Fassung EN 13306:2018-02, DIN Deutsches Institut für Normung. Berlin, 2018.
8 ifaa-Studie: Produktivitätsstrategien im Wandel – Digitalisierung in der deutschen Wirtschaft: Prof. Dr.-Ing. habil. Sascha Stowasser, 2019, Gefunden: 08.11.2024, 2019. www.arbeitswissenschaft.net/fileadmin/Downloads/Angebote_und_Produkte/Studien/ifaa-Studie_Produktivitaetsmanagement_2019.pdf.
9 Jacobs, Joh. Christian; Kagermann, Henning; Spath, Dieter (Hrsg.): Lebenslanges Lernen fördern – gute Beispiele aus der Praxis. Berlin: acatech – Deutsche Akademie der Technikwissenschaften. 2020, S. 22 ff.
10 Hirt, C.; Spahni, M.; Kompis, Y.; Jetter, D.; Kunz, A.: Alles rund um die Maschine – Begehbare virtuelle Schulung an Werkzeugmaschinen. In: VAR² 2019 – Realität erweitern, Hrsg. Putz, M.; Klimant, P.; Klimat, F.; Chemnitz, ISBN: 978-3-00-064420-7, S. 43–58, 2019.
11 Goldhahn, L.: Montageplanung und -ausführung – virtuell und real. 16th International Scientific Conference Mittweida IWKM. Scientific Reports Nr. 2, 2003, S. 46–50, 2003.
12 Klimant, P.; Klimant, F.; Putz, M.; Allmacher, C.: Trends der virtuellen und erweiterten Realität mit dem Fokus auf virtueller Inbetriebnahme. In: Reimund Neugebauer, Matthias Putz et Phillip Klimant, coord.: VAR² 2019 – Realität erweitern. Tagungsband. 5. Fachkonferenz zu VR/AR-Technologien in Anwendung und Forschung an der Professur Werkzeugmaschinen und Umformtechnik. Chemnitz, S. 7–26, 2019.
13 Eckardt, Robert; Goldhahn, Leif: Virtual Reality-basierte Bedienhandlungen zur Mitarbeiterqualifizierung am Beispiel einer Werkzeugmaschine. In: Nachwuchswissenschaftlerkonferenz 2020/21 der Ernst-Abbe-Hochschule Jena: 26./27.05.2021, ISBN-13: 978-3-932886-36-2, S. 79–83, 2021. www.eah-jena.de/fileadmin/user_upload/eah-jena.de/forschung/Promovenden/NWK/Tagungsband_NWK2021.pdf.
14 Engel, B.: Immersion oder Versinken in der virtuellen Realität – auch ein Thema für die Arbeitsmedizin? In: Zeitschrift für medizinische Prävention, Bd. 54, n° 09, S. 604–607. En ligne: 2019. www.asu-arbeitsmedizin.com/node/30413/print.

15 Thomas, O.; Metzger, D.; Niegemann, H. (Hrsg.): Digitalisierung in der Aus- und Weiterbildung. Berlin, Heidelberg: Springer, 2018.

16 Eckardt, Robert; Goldhahn, Leif: Virtual Reality-basierte Bedienhandlungen zur Mitarbeiterqualifizierung am Beispiel einer Werkzeugmaschine. In: Nachwuchswisse nschaftlerkonferenz 2020/21 der Ernst-Abbe-Hochschule Jena: 26./27.05.2021, ISBN-13: 978-3-932886-36-2, S. 79–83, 2021. www.eah-jena.de/fileadmin/user_upload/eah-jena.de/forschung/Promovenden/NWK/Tagungsband_NWK2021.pdf.

17 Hirt, C.; Spahni, M.; Kompis, Y.; Jetter, D.; Kunz, A.: Alles rund um die Maschine – Begehbare virtuelle Schulung an Werkzeugmaschinen. In: VAR² 2019 – Realität erweitern, Hrsg. Putz, M.; Klimant, P.; Klimat, F.; Chemnitz, ISBN: 978-3-00-064420-7, S. 43–58, 2019.

18 Wannemacher, K.; Jungermann, I.; Scholz, J.; Tercabli, H.; von Villiez, A.: Digitale Lernszenarien im Hochschulbereich, S. 16 Zugang 07.01.2020. https://hochschulforum digitalisierung.de/sites/default/files/dateien/HFD%20AP%20Nr%2015_Digitale%20Le rnszenarien.pdf (2016).

19 Schwan, S.; Buder, J.: Virtuelle Realität und E-Learning, S. 7–9, Zugang 07.01.2020, 2006. www.e-teaching.org/didaktik/gestaltung/vr/vr.pdf.

20 Eckardt, Robert; Goldhahn, Leif: Virtual Training for the Maintenance of Machine Tools. In: Isabel L. Nunes (eds.) Human Factors and Systems Interaction. AHFE (2024) International Conference. AHFE Open Access, vol. 154. AHFE International, USA, pp. 199–207, 2024. http://doi.org/10.54941/ahfe1005367.

21 VDI/VDE Richtlinie 3699 – Prozessführung mit Bildschirmen. Berlin, Beuth.

22 Zühlke, D.: Nutzergerechte Entwicklung von Mensch-Maschine-Systemen: Useware-Engineering für technische Systeme, 2. Aufl. Berlin: Springer, 2012.

23 DIN EN 13306: Instandhaltung – Begriffe der Instandhaltung; Dreisprachige Fassung EN 13306:2018-02, DIN Deutsches Institut für Normung. Berlin, Februar 2018.

24 Petersen, Inken: User Experience verstehen: In: Digitales Produktmanagement, Hoffmann, S. (Hrsg.) Springer Fachmedien Wiesbaden GmbH, ein Teil von Springer Nature, S. 327, 2023. https://doi.org/10.1007/978-3-658-41880-9_18.

25 Goldhahn, Leif; Eckard, Robert; Müller-Eppendorfer, Katharina: Standardisiertes Übungsszenario für HMD-basierte virtuelle Lernumgebungen. In: Ökologische Transformation in Technik, Wirtschaft und Gesellschaft? Scientific Reports. 26. Interdisziplinäre Wissenschaftliche Konferenz Mittweida, ISSN 1437-7624, Nr. 2, S. 105–108, 2021. https://doi.org/10.48446/opus-12300.

26 Klimant, P.; Klimant, F.; Putz, M.; Allmacher, C.: Trends der virtuellen und erweiterten Realität mit dem Fokus auf virtueller Inbetriebnahme. In: Reimund Neugebauer, Matthias Putz et Phillip Klimant, coord.: VAR² 2019 – Realität erweitern. Tagungsband. 5. Fachkonferenz zu VR/AR-Technologien in Anwendung und Forschung an der Professur Werkzeugmaschinen und Umformtechnik. Chemnitz, S. 7–26, 2019.

27 Eckardt, Robert; Goldhahn, Leif: Virtual Reality-basierte Bedienhandlungen zur Mitarbeiterqualifizierung am Beispiel einer Werkzeugmaschine. In: Nachwuchswisse nschaftlerkonferenz 2020/21 der Ernst-Abbe-Hochschule Jena: 26./27.05.2021, ISBN-13: 978-3-932886-36-2, S. 79–83, 2021. www.eah-jena.de/fileadmin/user_upload/eah-jena.de/forschung/Promovenden/NWK/Tagungsband_NWK2021.pdf.

28 Eckardt, Robert; Goldhahn, Leif: Virtual Reality-basierte Bedienhandlungen zur Mitarbeiterqualifizierung am Beispiel einer Werkzeugmaschine. In: Nachwuchswisse nschaftlerkonferenz 2020/21 der Ernst-Abbe-Hochschule Jena: 26./27.05.2021, ISBN-13: 978-3-932886-36-2, S. 79–83, 2021. www.eah-jena.de/fileadmin/user_upload/eah-jena.de/forschung/Promovenden/NWK/Tagungsband_NWK2021.pdf.

29 Goldhahn, L.; Eckardt, R.; Pietschmann, C.; Roch, S.: Lernszenarien und Virtual Reality-basierte Bedienhandlungen zur Mitarbeiterqualifizierung. In: Gesellschaft für Arbeitswissenschaft e. V. (Hrsg.): Digitaler Wandel, digitale Arbeit, digitaler Mensch? Bericht zum 66. Arbeitswissenschaftlichen Kongress vom 16.-18. März 2020. Dortmund: GfA Press, ISBN 978-3-936804-27-0, 2020, B1.1. S. 1–6, 2020.

30 Goldhahn, L.; Eckardt, R.; Pietschmann, C.; Roch, S.: Competency Profiles as a Means of Employee Advancement for a Resource-Efficient Chipping Production. In: Advances in Manufacturing, Production Management and Process Control, Proceedings of the AHFE 2019, 24–28.07.2019, Washington D.C., USA, pp. 146–157, 2019. https://doi.org/10.1007/978-3-030-20494-5_14.

31 Goldhahn et al. Competency Profiles as a Means of Employee Advancement.

32 Goldhahn et al. Competency Profiles as a Means of Employee Advancement.

33 VDI/VDE Richtlinie 3699 – Prozessführung mit Bildschirmen. Berlin, Beuth, 2015.

34 Hacker, W.: Allgemeine Arbeitspsychologie Psychische Regulation von Arbeitstätigkeiten, Huber Verlag, 1998.

35 Eckardt, Robert; Goldhahn, Leif: Virtual Reality-basierte Bedienhandlungen zur Mitarbeiterqualifizierung am Beispiel einer Werkzeugmaschine. In: Nachwuchswissenschaftlerkonferenz 2020/21 der Ernst-Abbe-Hochschule Jena: 26./27.05.2021, ISBN-13: 978-3-932886-36-2, S. 79–83, 2021. www.eah-jena.de/fileadmin/user_upload/eah-jena.de/forschung/Promovenden/NWK/Tagungsband_NWK2021.pdf.

36 Goldhahn, L.; Eckardt, R.; Pietschmann, C.; Roch, S.: Competency Profiles as a Means of Employee Advancement for a Resource-Efficient Chipping Production. In: Advances in Manufacturing, Production Management and Process Control, Proceedings of the AHFE 2019, 24–28.07. 2019, Washington D.C., USA, pp. 146–157, 2019. https://doi.org/10.1007/978-3-030-20494-5_14.

37 Goldhahn, L.; Eckardt, R.; Pietschmann, C.; Roch, S.: Lernszenarien und Virtual Reality-basierte Bedienhandlungen zur Mitarbeiterqualifizierung. In: Gesellschaft für Arbeitswissenschaft e. V. (Hrsg.): Digitaler Wandel, digitale Arbeit, digitaler Mensch? Bericht zum 66. Arbeitswissenschaftlichen Kongress vom 16.-18. März 2020. Dortmund: GfA Press, ISBN 978-3-936804-27-0, 2020, B1.1. S. 1–6, 2020.

38 Goldhahn, Leif; Bock, Dorit; Müller-Eppendorfer, Katharina: Virtuelle Lernszenarien für die manuelle Montage: Entwicklung und Validierung einer virtuellen Lernumgebung am Beispiel der Steckdosenmontage. Zeitschrift für wirtschaftlichen Fabrikbetrieb, Jahrg. 117, S. 763–767, 2022. https://doi.org/10.1515/zwf-2022-1145.

2 Wearables as an Easy, Cost-Effective and Smart Measurement Tool for Work System Evaluation

Fabian Dietrich, Stefanie Liebl, Marcus Löffler and Torsten Merkel

2.1 INTRODUCTION

In the modern world of work, which is characterized by rapid change, technological progress and increasing competitive pressure, the concept of "healthy work" is becoming increasingly important. The term "healthy work" encompasses not only the physical health of employees but also their mental well-being and social integration in the workplace. Companies are increasingly recognizing that promoting a health-conscious working environment not only increases the individual well-being of employees but also makes a decisive contribution to the long-term performance and competitiveness of the company. The challenges faced by employees in modern working environments – such as stress, time pressure and inadequate ergonomic conditions – can lead to physical and mental illness. It is therefore essential to take preventative measures and create an environment that promotes both the health and satisfaction of employees. In this regard, the implementation of ergonomic standards, the promotion of a positive corporate culture and the creation of opportunities for personal and professional development are particularly important. The promotion of healthy work is therefore an integral part of a sustainable corporate strategy.

Even if most company managers are aware of this fact, the resources required to carry out at least a basic analysis of working conditions and derive the necessary action and measures from this are often lacking in day-to-day operations. With the development of smart device technology such as wearables and smart measurement systems for recording the general conditions of work, the environment, points of movement in the room, etc., semi-automatic recording of work requirements and the associated stresses can be carried out over a longer period of time without the usual time-consuming and cost-intensive analysis concepts.

In the German project PAL (Perspektive Arbeitsforschung Lausitz), funded by the Federal Ministry of Education and Research, a sub-project is looking at how smart

DOI: 10.1201/9781003672036-2

measurement technologies can be used to enable small and medium-sized enterprises (SMEs) to carry out a basic analysis of working conditions without outside help and, with the evaluation options of a digital assistant, to obtain both the status of the quality of work and suggestions for optimizing requirements and working conditions. When defining more in-depth problems, reference is made to the necessary specialist expertise of third parties, whereby the upstream analysis already enables a strong focus on necessary priorities and the clear definition of a requirements profile.

2.2 STATE OF SCIENCE AND TECHNOLOGY: CLASSIC METHODS OF WORK SYSTEM EVALUATION

2.2.1 Healthy Work Screening (SGA)

The healthy working screening is a guideline-based, low-threshold orientation procedure in the form of an observation interview to identify and eliminate hazards in the company that are responsible for back complaints and illnesses and thus for particularly frequent absences. In addition, the "Screening Healthy Working" (SGA) guidelines are used to facilitate occupational reintegration (BEM) in accordance with Section 84 (2) SGB IX. The SGA is aimed at managers, occupational safety specialists, company doctors and employee representatives. The guidelines enable those involved in the company to implement occupational health and safety both on an ad hoc and preventative basis. At the same time, the statutory health and safety obligations are fulfilled. The SGA follows the seven steps agreed in the Joint German Occupational Health and Safety Strategy (GDA) for risk assessment in accordance with § 20a and § 20b of the Occupational Health and Safety Act. The guidelines were developed between 2004 and 2013 in collaboration with practicing physiotherapists and occupational psychologists from the Institute for Occupational, Organizational and Social Psychology at the Technical University of Dresden and the thematic initiative group "Body, Mind and Work – Holistic Prevention" based at INQA. On October 19, 2013, a clarification of the Occupational Health and Safety Act (ArbSchG) came into force in Germany. As a result, employers are obliged to ensure both the physical and mental health of employees in the workplace in accordance with Section 4, No. 1 ArbSchG.[1] The SGA is fundamentally applicable in all industries and can be reliably applied by any person within a relatively short time using the guidelines provided in combination with the published knowledge base.

The assessment is based on various criteria, which can be divided into the following categories:

1. **Employment relationship:** employment contract, place of work and working hours
2. **Physical strain:**
   a. **Sitting:** duration of sitting, ergonomics of sitting
   b. **Standing:** duration of standing, ergonomics of standing
   c. **Walking:** walking and standing, transportation

   d. **Work equipment:** use of work equipment, posture, implementation, mechanical oscillations/vibration, activities involving load handling, repetitive movements, range of motion, working close to the body, working heights
3. **Mental stress:**
   a. **Work activity:** task variety and execution, work intensity, holistic nature of the activity, repetition, scope of activity, freedom from contradictions, feedback culture, availability of information, customer contact
   b. **Work organization:** work process, responsibility, cooperation, stability of cooperation, participation, short breaks
   c. **Social conditions:** social support, management style, recognition
4. **Working environment:** lighting, noise, odor, indoor climate/draft, etc.

In addition, interviews were conducted with the previously observed employees in order to gain deeper insights into their individual experiences and perceptions. The qualitative component of the study enabled the researchers to identify specific problem areas that were not adequately captured through observation. The time required to conduct the observation interviews varied between 30 and 180 minutes, depending on the workplace and situation.

The observation forms are evaluated by means of a simple count of the identified hazards. A hazard exists if a plus is noted next to the "yes/no" answer in the evaluation form (see Figure 2.1). Counting the answers marked with a plus results in an evaluation scheme (see Figure 2.2). If the values determined are in the area marked in red, it can be assumed that there is a serious risk at the workplace that requires immediate action. If the values are in the yellow area, there are minor deficiencies whose rectification is recommended but not absolutely necessary, while the green area indicates an (almost) ideal workplace.

2.2.1 Key Feature Methods

The guiding characteristic methods published by the Federal Institute for Occupational Safety and Health have developed into a standard as the basic instrument for the assessment of physical stress, in particular lifting, carrying, pulling, pushing and other manual work.[2] These methods are already available in an electronic version, which greatly simplifies both the determination of the load situation and the traceability of possible design approaches. However, due to the nature of the procedure, comprehensive observation of the workplace and all associated physical stresses and their exact measurement are still necessary for the exact recording of the stresses that occur.

2.3 VITAL DATA RECORDING WITH ADHESIVE ELECTRODES

Cardiac activity is usually examined by means of an electrocardiogram (ECG). This procedure requires the recording of the electrical voltage that occurs during the

Examiner: SGA observation sheet
Date:

Working area: | Numer of employees: | (Partial) activity:

Box with +: low or no risk to health | Box without +: There is a risk to health

A) Employment relationship

Feature	Description	Yes	No	Knowledge repository
EMPLOYMENT RELATIONSHIP	**How is the employment contract structured?**			
Labour contract	**A1** Fixed-term employment contract?	☐	☐+	- If at least one of the characteristics applies, incorrect utilisation is to be expected
Place of Work	**A2** Temporary labour/temporary work?	☐	☐+	- Exception: 'Yes' for 'Is it freelance work' not automatically a risk, it must be questioned whether this is unsafe or beneficial
Working hours	**A3** Are they freelancers?	☐	☐+	- If the characteristics cannot be changed due to operational circumstances, the optimisation of the following factors relating to physical and mental stress and occupational hygiene is given higher priority
	A4 Do you work at a fixed location/area?	☐+	☐	
	A5 Do you regularly have to work night shifts?	☐	☐+	**- Recommendation: Clarify questions about the contract in advance**
	A6 Are on-call services to be provided?	☐	☐+	
	A7 Do you regularly work overtime (more than 48 hours per week)?	☐	☐+	

FIGURE 2.1 Excerpt from the Healthy Work Screening observation form.

Excerpt from the evaluation criteria of the SGA questionnaire including description and entry from the knowledge store.

Source: **Own presentation. Following Debitz et al. (2016).**

contraction of the heart muscle. This is achieved by attaching adhesive electrodes to the skin in the chest area. The measured values are generally highly accurate, but the wearing comfort and freedom of movement are suboptimal due to the necessary cabling and the application of the electrodes directly to the skin. Furthermore, the adhesion of the measuring equipment is sometimes limited, especially if the test subject is heavily perspiring. As a result, the recording of an ECG (e.g., in phases of stress or rest) is often used in a medical environment to localize specific health problems. In contrast, they are rarely used in the field of occupational science in situ investigations at the workplace due to the obstacles described. Exceptions are clearly defined laboratory tests, in which the negative aspects are accepted in favor of high data quality.

2.4 MEASUREMENT OF ENVIRONMENTAL FACTORS

A classic element of work system evaluation is the measurement of work environment factors. As these are physical variables, there are a large number of measuring devices

Factor	Scale
Employment relationships Fixed-term contracts, temporary agency work/temporary work, fixed place of work, shift work, on-call duty, overtime	(0) 1 2 3 4 5 6
Physical stress	
Posture Sitting, standing, walking	0 1 2 (3) 4 5 6
Work equipment Transport, mechanical oscillations + vibration	0 (1) 2 3 4 5
Ergonomics Load handling, repetitive movements, range of motion, working close to the body, working heights	0 1 2 (3) 4 5
Mental stress	
Work activity Variety of tasks, work intensity, completeness, repetition, scope of activities, consistency, feedback, information, customer contact	0 1 (2) 3 4 5 6 7 8 9 (range marked 1–3)
Work organization Procedure, responsibility, cooperation, co-determination, short breaks	0 (1) 2 3 4 5 6
Social relationships Support, management style, recognition	(0) 1 2 3
Working environment Lighting, noise, odor, indoor climate	0 1 2 (3) 4

FIGURE 2.2 Evaluation sheet for screening healthy work in the example company in the assembly area.

Table with the evaluation factors described in the text. Low risk values including standard deviation are entered in all categories.

Source: **Own representation. Following Debitz et al. (2016).**

of varying complexity, accuracy and cost. High-quality work environment measurement technology not only determines the data of the relevant environmental factors but also provides direct information on whether legal or normative limit values have been exceeded.[3]

2.5 ADVANTAGES AND DISADVANTAGES OF TRADITIONAL RECORDING METHODS

The measurement methods described are chosen as the preferred solution for comprehensive work system evaluation in companies, primarily due to the accuracy of the data collected and the validated results. Due to the high complexity of data collection and, above all, evaluation, the users are usually not the company's employees but external specialists in occupational science or occupational medicine. This relieves companies of the responsibility for data collection and evaluation. On the other hand, outsourcing these necessary processes leads to high costs. At the same time, it should be emphasized that the majority of workplaces in Germany fortunately have a very high standard in terms of employee well-being. As a result, companies may make a large investment only to find that there is no or only minimal need for action. As a result, many companies, especially SMEs, avoid a generalized examination of work systems and only have the minimum of checks required by law carried out in order to reduce expenditure. As a consequence, suboptimally designed workplaces and processes often remain undetected.

2.6 NEW APPROACH: TREND ANALYSIS USING WEARABLES

The use of wearables offers advantages when conducting long-term studies on the holistic effects of work. This approach is attractive both in terms of the investment in the device technology and the costs of specialist support. While the selective recordings with qualified measurement technology can only capture the image of the situation found, wearables in combination with smart measurement technology for the working environment create an image of cyclical changes over longer periods of time (see Figure 2.3). By linking the vital data from the wearables with issues relating to work severity and the working environment, the physical and psychological effects on staff can also be understood, and suitable measures can be derived.

2.7 COMPONENTS OF THE SMART TREND SURVEY

2.7.1 Optical Pulse Sensor

Polar verity sense was used as the optical pulse sensor (photo-plethysmography principle) in the tests and studies carried out to date. In principle, any other sensor that records the heart rate, makes the data available for export via an interface and can meet the other requirements of the application-specific specifications is also suitable. This is the case with the selected sensor:

- Low-energy Bluetooth connection
- Internal memory for recording data during disconnections

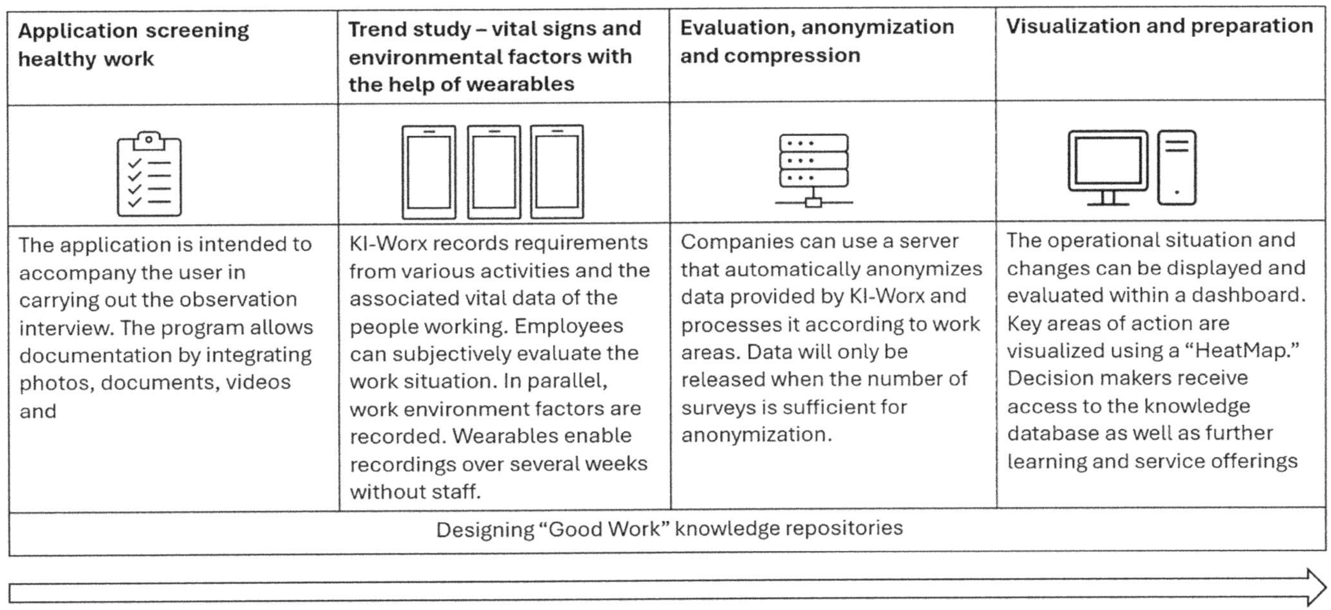

FIGURE 2.3 Schematic process of a smart work system evaluation in the company.

Process of the study using vital signs evaluation. 1. SGA, 2. Recording of measured values, 3. Evaluation of the data, 4. Presentation and presentation of the results and suggested solutions.

- No data transfer to cloud or software from external providers necessary (data protection)
- Long battery life (>24 h)
- Small form factor for unobstructed wearing of the sensor
- Low price (<100 €)

2.7.2 KIWorx

A mobile application called KIWorx was specially developed at the University of Applied Sciences Zwickau as part of the PAL project to store and process the data collected by the pulse sensor. This is installed on a smartphone connected to the sensor via Bluetooth and does not require internet access. The heart rate data is only stored locally at a defined time interval (e.g., 5 s). The Baevsky Stress Index and the measure of the employee's ability to recover are also calculated based on the measured heart rate variability.[4] These trend values are displayed to the user as an overview diagram and by means of clearly understandable traffic light values (see Figure 2.4).

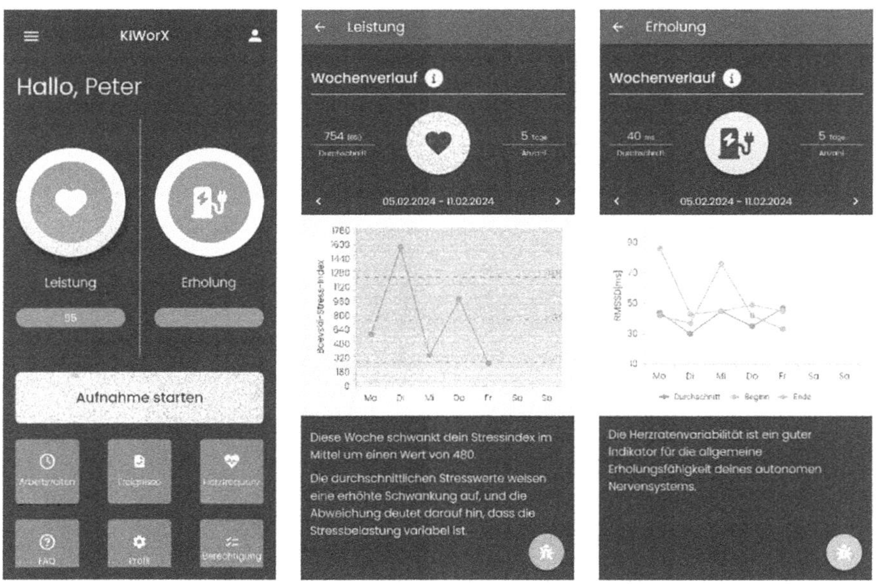

FIGURE 2.4 Displays in the KIWorx app: performance and recovery status (left), Baevsky Stress Index (BSI) over the course of the week (middle) and heart rate variability over the course of the week (right).

Screenshots from the KIWorx app: traffic light for performance and recovery (left), diagram of the BSI weekly progression (middle) and diagram of the HRV weekly progression (right).

These visual representations are not fundamentally necessary for the functionality of the system. However, in order to offer employees added value and allow them to participate in the recording of measured values, a conscious decision was made to prepare the data accordingly. In parallel tests, test subjects were provided with a similar system from the Apple system environment (iPhone with Apple Health in combination with Apple Watch). The employees expressed negative opinions about the lack of traceability of the specific data measured and relevant to the study.

Furthermore, the KIWorx app integrates the manual recording of workplaces and activities by the employee for a concrete evaluation as well as the query of subjective perceptions, for example, by inserting the NASA Task Load Index (TLX)[5] questionnaire, based on a diary study. In addition to the objective recording of measured values, this also enables specific psychological and individual factors to be considered.

2.7.3 Environmental Measuring Device

An Air-Q pro is used as a smart environmental measuring device. In contrast to comparable devices with dedicated use for the evaluation of environmental factors in the workplace (e.g., source), this is handy (13.50 × 4.90 × 11.80 cm; 0.4 kg), very robust and, at only approx. 15% of the cost, also significantly cheaper. Thirteen integrated sensors measure the relevant environmental factors with sufficient accuracy to reliably detect when legally prescribed or occupational health recommended limit values are exceeded. Furthermore, the recorded values are displayed online in a browser interface without additional software installation; the interface can also be operated and interpreted intuitively by non-experts.

2.7.4 Dashboard

The dashboard as an evaluation and display application is provided as an executable file for computers and is based on web technologies, displayed in a browser. Development was carried out using Python 3.9.13, utilizing the Dash and Plotly libraries. The dashboard is designed to process CSV files from various data sources, which can be uploaded via a drag-and-drop menu. It displays both environmental and physiological parameters. Table 2.1 lists all the parameters that can currently be analyzed and displayed. Users can input additional information, such as group names

TABLE 2.1
Selectable Parameters in the Dashboard

Physiological Parameter	Environmental Parameter
Heart rate	Carbon dioxide content in the air
Inter-beat intervals (IBIs)	Oxygen content in the air
Root mean square of IBIs (RMSSD)	Sound
Standard deviation of IBIs (SDNN)	Temperature
Stress index according to Baevsky (BSI)	Particle content in the air

or work sections, to facilitate comparisons. Uploaded data can be selected via a dropdown menu, and areas for analysis can be chosen using the mouse. The dashboard also provides recommendations based on current literature. For instance, it can flag thresholds for high workload or elevated concentrations of carbon dioxide in the air. The KIWorX app enables the recording of work tasks, which can be uploaded to the dashboard with timestamps. This feature allows data analysis to be linked to specific tasks in an industrial environment. The dashboard is segmented into three areas, as shown in Figure 2.5. Area (a) contains the inputs and dropdown menus for selecting data and parameters to be analyzed and displayed. Area (b) contains the main graph on which all data is shown. Specific data analysis, such as recommendations or comparisons between departments, is done in area (c). For example, if the inter-beat intervals (IBIs) are selected in area (b), it is possible to create a poincaré plot to analyze the distribution of the IBIs. If the root mean square of IBIs (RMSSD) is selected, the user can analyze what percentage of the values falls within defined normal domains. Created plots can be exported as figure.

2.8 ADVANTAGES AND DISADVANTAGES OF THE NEW, SMART RECORDING METHOD

The advantages of the smart, mobile technologies currently used in measurement technology lie in the possibility of carrying out longer studies in the company under its own responsibility and care without the permanent involvement of specialist personnel. As a rule, an inspection of the work area is sufficient to determine possible positions for environmental measurement technology or "beacons" for determining mobility requirements. The wearables worn on the body are self-explanatory. The main task of the occupational scientist on site is merely to provide the necessary background information for further processing of the data and the charging cycles required for such systems, as well as the subsequent optimization of the main points of action identified by the system in cooperation with the company. The systems are also suitable for determining trends over several days or on a weekly or quarterly basis. This makes it possible to track changes in climate quality or exposure to hazardous substances over the course of a day or even a year. The ability to recover can also be determined over the working week or month. The total costs for such a system are significantly lower than the (temporary) employment of a qualified occupational scientist. Furthermore, in addition to legally prescribed and thus to a certain extent generalized limit values, subjective characteristics and predispositions are also considered in the system described. In this way, work systems can be fully adapted to the individual employee, or the employee can be transferred to a different workplace that is more suitable for them based on their individual physical and psychological characteristics on the basis of objective measurement data.

The disadvantages of the devices include significantly poorer measurement quality. In particular, the heart rate measurements of wearables based on the photochemical detection of the pulse wave are prone to errors. Unforeseeable or unexpected events can only be recorded retrospectively. If no analysis personnel are nearby, important information is lost, and the measurements may have to be repeated.

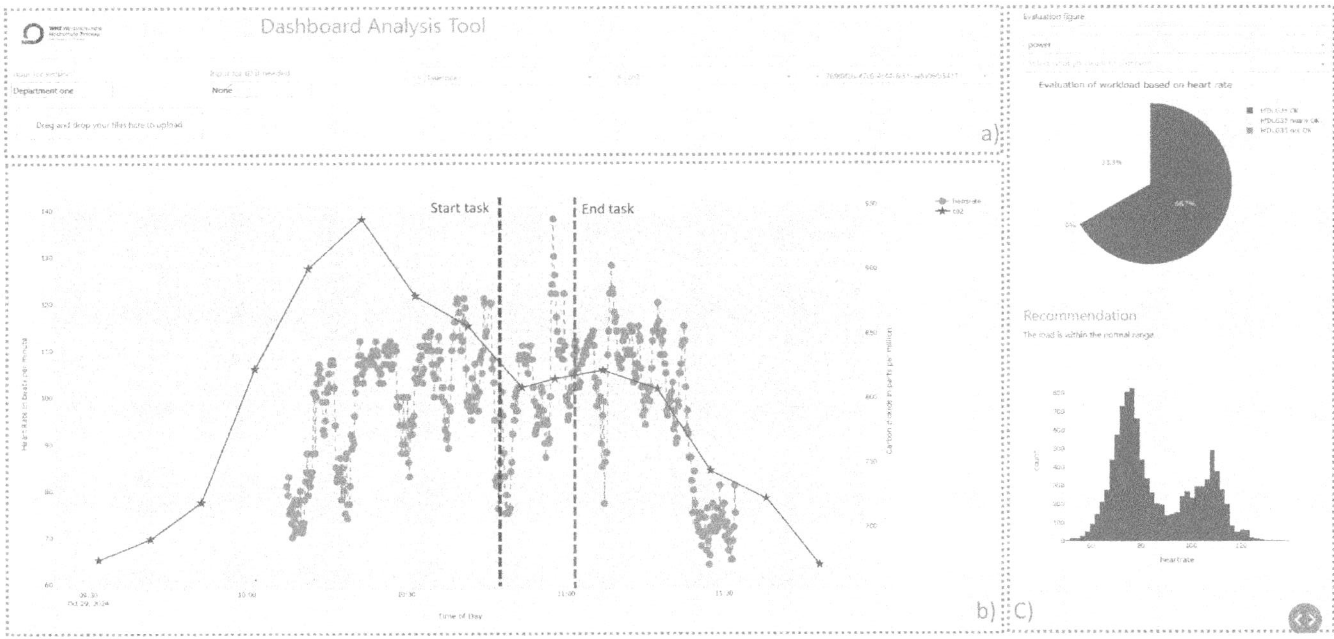

FIGURE 2.5 (a) The inputs and dropdown menus. (b) Area for data visualization. (c) Section where different types of data evaluation can be selected. In the upper left part there is the selection field of the dashboard and below it is a line or scatter diagram with the course of the CO_2 concentration and the heart rate. In the upper right part there is a pie chart that shows the frequency distribution of the heart rate with respect to the applicable limit values in traffic light colors (33.3% yellow, 67.7% green). Recommendations for workplace optimization can be displayed below, and the heart rate distribution can be seen in a bar chart below.

In order to guarantee data protection, systems were selected that also function without the device manufacturer's cloud solution. This procedure has the effect that every interface change has to be entered into the self-created applications. In addition, corrections and evaluations that the manufacturers of wearables generate in their cloud applications are not available. Corresponding adaptation and optimization of algorithms must be carried out by the working group itself. This is time-consuming and has a high potential for error. However, the procedure opens up the possibility of linking different data sources and, instead of frequently sport and fitness-oriented evaluations, actually carrying out a holistic work-oriented optimization.

2.9 FEASIBILITY STUDY

First Applications of Wearables as Easy, Cost-Effective and Smart Measurement Tool for Work System Evaluation

For an initial test of the applicability of wearables for recording vital data in an operational context, a trial by the Faculty of Health and Nursing Sciences at the West Saxon University of Applied Sciences in Zwickau was equipped with measuring systems on the subject of "Motivation in digital work." The aim of the study was to investigate age-specific differences in motivation and stress levels when working on complex office tasks. The cognitive demands of screen work and digital forms of communication were examined, which can often lead to a loss of motivation and performance as well as health problems. The work task was operationalized using the "Job-related test of cognitive competence" (SMART), while motivation was measured using the questionnaire for recording current motivation in learning and performance situations (FAM). The central question is how external digital interruptions and emotions influence current motivation as a function of age.

The experimental procedure is based on the study by Foroughi et al.,[6] which shows that interruptions have a negative impact on the quality of work. In a pre-test, the processes and technology were tested with four female employees at the university. For the experiment, a standardized test procedure (SMART) was acquired via the Viennese test system Schuhfried. The test measures cognitive abilities such as logical reasoning and verbal and numerical skills. The experiment involved a rest period followed by completion of the SMART test under two conditions: an experimental group with four 60-second digital interruptions that included an additional cognitive task and a control group that had to complete the tasks without interruptions. The tests took place in a specially equipped office workstation at the West Saxon University of Applied Sciences Zwickau. After the test, the test subjects received feedback on their results.

In addition to evaluating the questionnaires and work tasks, the test subjects were also equipped with two measuring systems for measuring vital signs:

- Polar OH1 with smartphone incl. KIWorx application
- Movisense ECG sensor with two adhesive electrodes in the chest area

Before the start of the study, the sensors were handed over to the test subjects by a member of staff from the Faculty of Occupational Sciences, checked for correct fit after being put on and the measurement initialized. The aim of using both measurement systems was to determine the deviation and thus the significance of the wearables solution in comparison to a highly precise and established ECG measurement.

The health science investigations of the study showed that the interruption tasks did not lead to a measurably higher stress level (see Figure 2.6) and, analogously, that the motivation and confidence to successfully complete the set tasks increased. It can therefore be concluded that the design of the selected interruption was too conservative and/or that the group of test subjects, which primarily consisted of students and university staff, was too familiar with this type of interruption for it to cause stress.

Furthermore, the reliability and deviation of the two measuring systems from each other were determined. As shown in Figure 2.7, the Polar sensor delivers a consistent signal. By calculating the mean, we find that the difference is small, with a value of −0.038 BPM (beats per minute). However, due to the averaged nature of the signal, the absolute mean is calculated to be 2.62 BPM. The t-test results, with a t-statistic of −0.741 and a p-value of 0.459, indicate that there is no significant difference between the two datasets.

In summary, the investigations carried out demonstrated the fundamental applicability of wearables for work system evaluation in the laboratory environment. Furthermore, the present study opens up the possibility of use in an operational environment due to minor ergonomic restrictions and good measurement data acquisition and quality of the measurement system.

2.10 FIRST OPERATIONAL APPLICATION

First Applications of Wearables as Easy, Cost-Effective and Smart Measurement Tool for Work System Evaluation

A test was carried out at a partner company from the PAL project to verify the applicability of the wearables in an operational environment. The objectives were

- Comparison of the wearable measurement data with the results of the classic workplace evaluation
- Query the usability experience as a basis for optimizing the overall system

The subject of the investigations was a small department with ten employees, which ensures the remote operation of handling and production systems in external companies. Each employee in the shared office operates four screens and a laptop to monitor several machines, ensure their smooth operation and, if necessary, provide employees on site with work instructions for troubleshooting. This work requires a high level of technical expertise and entails the risk of persistently high stress levels due to the constantly high level of responsibility.

To analyze the actual stress levels in the department in question, an observation interview (SGA) was conducted with each employee present (6 including the head of department), supported by a measurement of the environmental parameters using

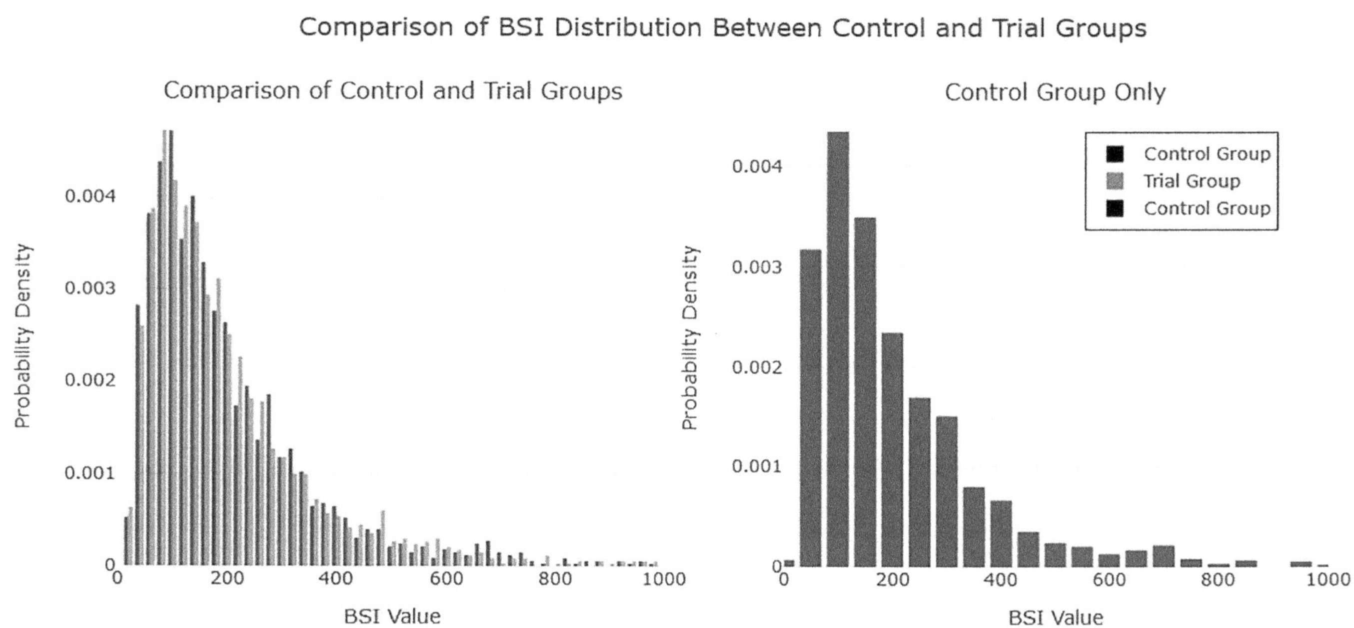

FIGURE 2.6 Comparison of the BSI distribution between the control and trial groups.
Diagram of the BSI distribution, maximum of the course at approx. 125, course and probability distribution in the experimental group and control group almost identical.

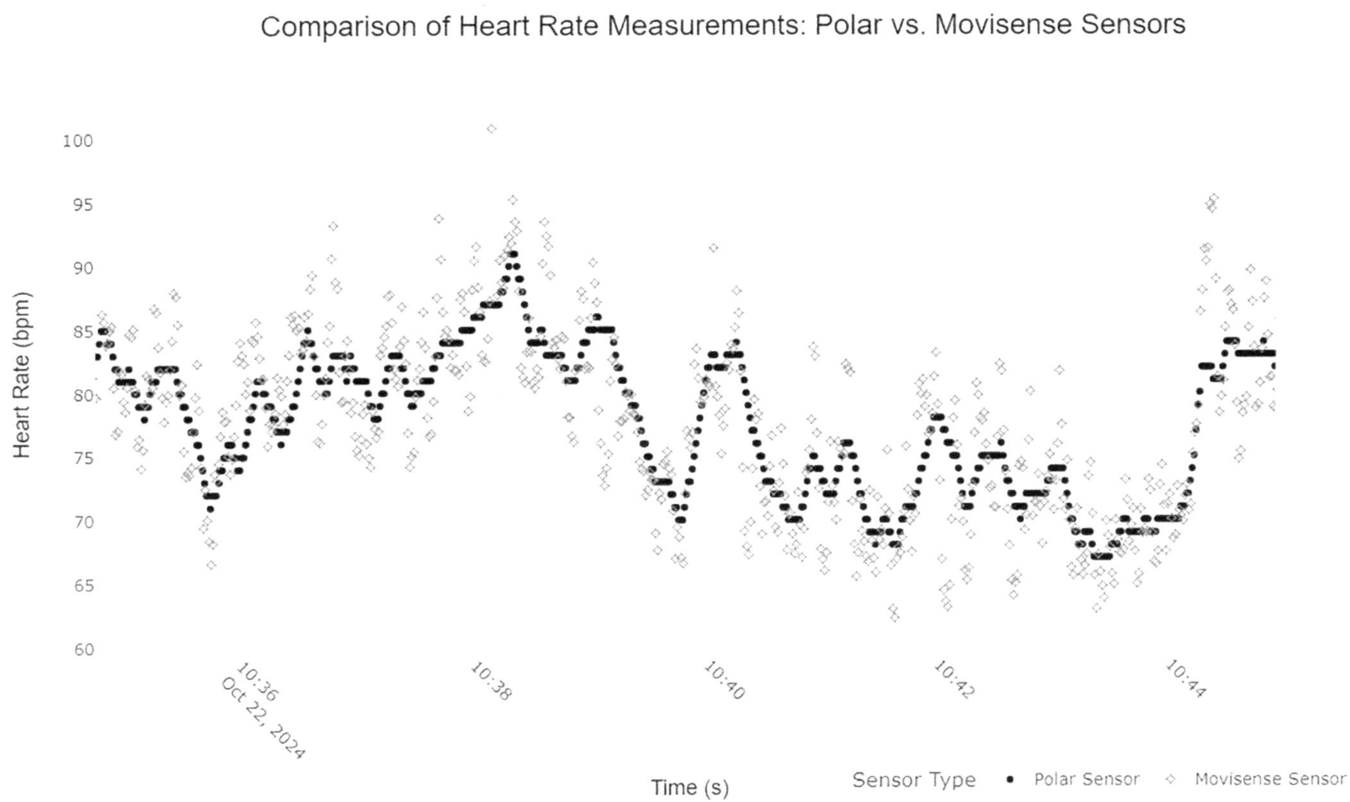

FIGURE 2.7 Comparison of heart rate measurements: Photoplethysmography (dots) and adhesive electrodes (circles).
Graph showing HRV history during an office experiment. Values of the wearable with less significant fluctuations, but average values almost identical.

a Bappu Evo. The results show that the climatic conditions in the office could be optimized but are not yet a cause for concern. Furthermore, noise was identified as a disruptive factor, both subjectively perceived by the employees and objectively measured.

During the assembly process, several minor deficiencies were identified that could affect both the physical and mental health of employees. The analysis carried out leads to the conclusion that the ergonomic conditions at the workplace are inadequate, which results in an increase in mental stress. In this example, the deficiencies in assembly identified in the analysis only require minor measures to improve ergonomic conditions and reduce mental stress. Targeted adjustments to work equipment, regular feedback and optimized work organization can increase the well-being of employees and promote their productivity in the long term.

The data obtained using wearables confirms the trends previously determined using SGA. No measurement days with significant stressors can be identified. Instead, the values show a trend of good utilization of the employees surveyed with slight excesses in the direction of increased stress (BSI 215). The measured values are therefore far from exceeding the limit value of 400 (see Figure 2.8).

If the investigations had been carried out as part of the company's own internal initiative, it would not necessarily have been necessary to focus on or optimize the given workplaces. However, both the results of the SGA and the measured values of the wearables indicate that an optimization of the examined workplaces is possible and should be considered reasonable.

2.11 SUMMARY AND OUTLOOK

Initial tests in the laboratory environment and in an operational context have shown that identifying key areas for action as part of work system evaluation is a promising concept. In contrast to the use of classic examination systems, the system can be used within the company after a short training period so that a comprehensive evaluation of all workplaces and processes can be realized with a relatively low investment. This enables small and medium-sized companies in particular to create optimum working conditions for their employees. Furthermore, with the exception of the recording of environmental measurement data, the system functions completely offline and autonomously, which is why, in combination with the anonymization and pseudonymization of the measurement data, the current data protection requirements are also met. This also helps to allay employees' concerns regarding the evaluation of their personal performance and behavioral patterns in the workplace.

A partner company of the PAL project in Lusatia is currently conducting a further test under real conditions with 20 wearable measurement systems in two departments (development/design and assembly). The data obtained from this will be compared with values determined in advance using conventional measurement methods. In this way, the interpretation of the values recorded using wearables can be made more precise.

In the future, it is conceivable that the measurement system could be expanded to include additional sensor elements. For example, the functionality of the KIWorx

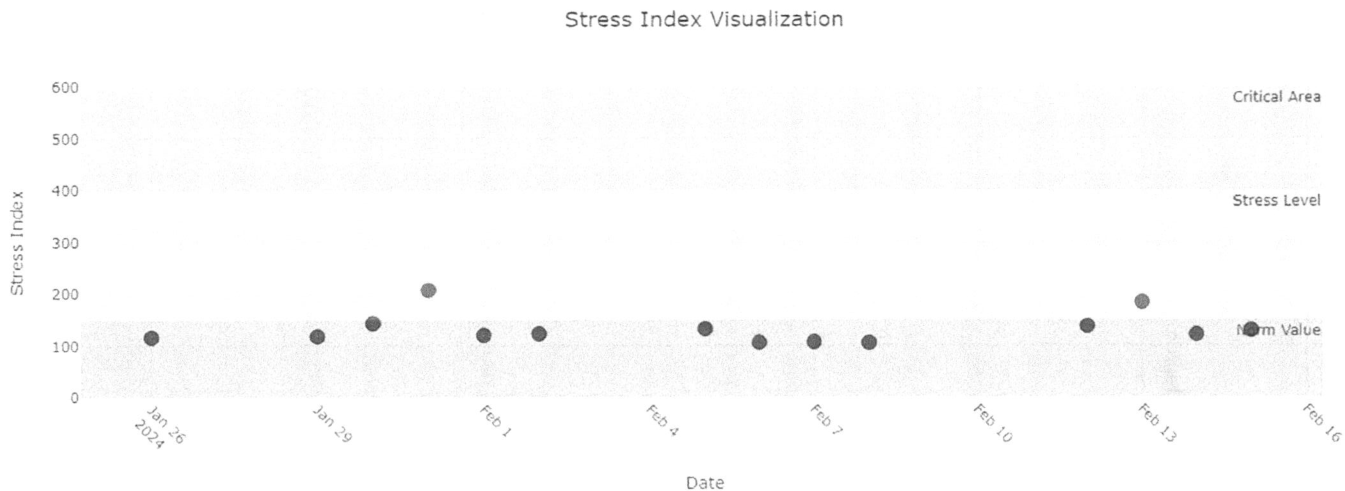

FIGURE 2.8 Determined, averaged BSI of the entire department examined.
Measuring points of daily BSI values in a diagram, which is colored in traffic light colors based on the limit values. Out of 14 points, 12 are in the upper green range, while 2 points are in the low yellow range.

app can be supplemented by integrating sensor soles. These can be used to determine ground reaction forces, which in turn allow conclusions to be drawn about the type and extent of moving masses and movement sequences as a counterpart to the leading feature method.

NOTES

1 Debitz, U., G. Buruck, S. Mühlpfordt, E. Muzykorska, U. Lübbert and H. Schmidt, *Der Leitfaden zum Screening Gesundes Arbeiten – SGA*. Initiative Neue Qualität der Arbeit, 2016.

2 Bundesanstalt für Arbeitsschutz und Arbeitsmedizin (BAuA). *Leitmerkmalmethode*. Accessed December 8, 2024. www.baua.de/DE/Themen/Arbeitsgestaltung/Gefaehrdungsbeurteilung/.

3 ELK-Gesellschaft für Erstellung, Layout und Konzeption elektronischer Systeme mbH. *Bappu: Komponenten*. Accessed October 8, 2024. www.bappu.de/de/komponenten/bappu.

4 Baevsky, R. M. and A. G. Chernikova, "Heart Rate Variability Analysis: Physiological Foundations and Main Methods." *Cardiometry*. 2017. https://doi.org/10.12710/cardiometry.2017.10.6676.

5 NASA Task Load Index (TLX). "TLX Scale." Accessed December 8, 2024. https://humansystems.arc.nasa.gov/groups/tlx/downloads/TLXScale.pdf.

6 Foroughi, C., N. E. Werner, E. T. Nelson and D. A. Boehm-Davis. "Do Interruptions Affect Quality of Work?" *Human Factors* 56, no. 7: 1262–1271. 2014. https://doi.org/10.1177/0018720814531786.

REFERENCES

Baevsky, R. M. and A. G. Chernikova, "Heart Rate Variability Analysis: Physiological Foundations and Main Methods." *Cardiometry* 10: 66–76. 2017. https://doi.org/10.12710/cardiometry.2017.10.6676.

Bundesanstalt für Arbeitsschutz und Arbeitsmedizin (BAuA). *Leitmerkmalmethode*. Accessed December 8, 2024. www.baua.de/DE/Themen/Arbeitsgestaltung/Gefaehrdungsbeurteilung/.

Debitz, U., G. Buruck, S. Mühlpfordt, E. Muzykorska, U. Lübbert and H. Schmidt, *Der Leitfaden zum Screening Gesundes Arbeiten – SGA*. Initiative Neue Qualität der Arbeit, 2016.

ELK-Gesellschaft für Erstellung, Layout und Konzeption elektronischer Systeme mbH. *Bappu: Komponenten*. Accessed October 8, 2024. www.bappu.de/de/komponenten/bappu.

Foroughi, C., N. E. Werner, E. T. Nelson and D. A. Boehm-Davis. "Do Interruptions Affect Quality of Work?" *Human Factors* 56, no. 7: 1262–1271. 2014. https://doi.org/10.1177/0018720814531786.

NASA Task Load Index (TLX). "TLX Scale." Accessed December 8, 2024. https://humansystems.arc.nasa.gov/groups/tlx/downloads/TLXScale.pdf

3 Wearable Technology for Lower Limbs Ergonomic Risk Assessment of Surgeons

Catarina Santos, Ana Teresa Gabriel, Cláudia Quaresma and Isabel L. Nunes

3.1 INTRODUCTION

Work-related musculoskeletal disorders (WRMSD) encompass a wide range of inflammatory and degenerative conditions that are induced and/or exacerbated by work-related physical, psychosocial, and individual risk factors (De Kok et al., 2019; Nunes, 2009). These conditions impact the muscles, tendons, ligaments, joints, peripheral nerves, and supporting blood vessels, representing a major occupational health concern (De Kok et al., 2019).

In recent years, there has been growing recognition in the surgical field regarding WRMSD and their contributing risk factors. Research has identified a high prevalence of WRMSD among surgeons, highlighting the importance of addressing both WRMSD and their associated risk factors (Epstein et al., 2018).

Recent studies show that surgeons are susceptible to the development of WRMSD that affect the upper limbs, back, and lower limbs (Epstein et al., 2018; Santos, Gabriel, Quaresma, & Nunes, 2024c; Swank et al., 2022; Xu et al., 2023). Nevertheless, the discussion surrounding WRMSD of the lower limbs and its associated symptoms remains sparse and fragmented throughout the scientific literature (Santos, Gabriel, Quaresma, & Nunes, 2024c; Xu et al., 2023; AlQahtani et al., 2016; Alqahtani et al., 2016). This knowledge gap has constrained a thorough understanding of the prevalence, impact, and contributing risk factors of these conditions among surgeons.

Despite this, research has shown that prolonged standing and maintaining upright postures, particularly during lengthy surgical procedures, are significant risk factors for developing lower limb WRMSD (Santos, Gabriel, Quaresma, & Nunes, 2024c; AlQahtani et al., 2016; Alqahtani et al., 2016). Furthermore, studies about other occupations that require prolonged standing have further demonstrated the significance of standing duration, working posture, muscle exertion, and holding time conditions as important risk factors in the development of lower limb WRMSD (Aitchison et al., 2016; Awosan et al., 2017; Halim & Omar, 2012).

DOI: 10.1201/9781003672036-3

It is also important to recognize the potential negative impact of lower limb WRMSD on mobility and balance, which can compromise both the surgeon's health and the precision required in surgical procedures, potentially affecting patient surgery outcomes and the efficiency of the healthcare system (Alzahrani et al., 2016; Swank et al., 2022). Given the negative consequences posed by lower limb WRMSD, there is an urgent need for ergonomic risk assessment tools specifically to evaluate the risk factors contributing to the development of lower limb WRMSD affecting surgeons. Such efforts are crucial for improving surgeon's well-being, enhancing the patient care quality, and ensuring the long-term sustainability of the healthcare workforce.

Despite the evident need, current ergonomic risk assessment tools are limited in their ability to specifically evaluate lower limb risks. Most tools, including lower limb evaluation, are designed for full-body assessments and provide only limited attention to the lower limbs. Furthermore, data collection methods often rely on self-reporting and observational approaches, which involve time-consuming manual data collection processes and lack the objectivity required to assess risk factor's exposure effectively (David, 2005).

Advancements in wearable technology present a promising solution for real-time, quantitative approach to monitor surgeon's exposure to risk factors during surgeries (Santos, Gabriel, Quaresma, & Nunes, 2024a). These small, lightweight devices, equipped with powerful sensing, processing, storage, and data transmission capabilities, continuously capture a range of activity data, including movement and biometric information (Cheng & Mitomo, 2017; Jeong et al., 2017; Koutromanos & Kazakou, 2020).

Integrating wearable sensors into ergonomic risk assessment methods for surgical procedures has the potential to greatly enhance the identification and prevention of WRMSD in the operating rooms. These sensors provide continuous and objective data, which can revolutionize approaches to ergonomic risk assessment, enhance surgeon well-being, and lead to improved surgical efficiency.

This work aims to synthesize the evidence on the use of wearable sensors for lower limb ergonomic risk assessment of surgeons performing surgeries. It includes a comprehensive analysis of the assessments outlined in the studies.

3.2 METHODOLOGY

The methodology adopted in this paper is a systematic literature review. This approach is designed to assess, interpret, and select the relevant studies based on predetermined inclusion criteria that address the research objectives (Xiao & Watson, 2017).

The systematic review follows the preferred reporting items for systematic reviews and meta-analyses (PRISMA) guidelines (Page et al., 2021). These guidelines provided a framework for outlining the eligibility criteria, search strategy, and information elicitation process.

The adoption of PRISMA ensures that this systematic review maintains methodological rigor, allowing for an objective synthesis of evidence that can support future studies (Kitchenham et al., 2010; Kitchenham & Charters, 2007).

3.2.1 Eligibility Criteria

A set of inclusion and exclusion criteria was established to refine the search. The relevance of the papers was evaluated using the following inclusion criteria:

- Papers written in English;
- Only papers published in scientific journals, conference papers, or proceedings;
- Studies describing the use of wearable sensors to monitor surgeon's lower limbs for ergonomic risk assessment purposes during surgical procedures.

The exclusion criteria were as follows:

- Review papers;
- Papers that focus on using wearable sensors for ergonomic risk assessments in other occupations, such as nursing;
- Papers that focus on data collection protocols, providing only predictions of potential outcomes within a work environment;
- Studies where data collected from wearable sensors are not used to conduct an ergonomic risk assessment of surgeons.

3.2.2 Search Strategy

The search was conducted in September 2024 using Scopus, PubMed, and Web of Science databases. To align with this systematic review's objectives, the following search string was used: (surgeon) AND (wearable OR sensor) AND (ergonomic*). The search was limited to articles, conference papers, and proceedings, except for PubMed, where these filters are unavailable. To maximize the number of results, the "All Fields" option was applied to all terms across the three databases.

An exploratory search string that included terms like "lower limb," "hip," "knee," and "ankle" was also tested. However, some relevant studies did not use these terms in their titles, abstracts, or keywords, despite addressing wearable sensor use on lower limbs for ergonomic risk assessments of surgeons. Since Scopus and Web of Science restrict searches to titles, abstracts, and keywords only, the lower limb-specific terms were removed from the final search string to avoid overlooking relevant studies.

Once the papers were retrieved from the databases, they were screened based on the defined eligibility criteria. Initially, the papers not meeting the inclusion criteria were excluded based on their titles and abstracts. The remaining papers were reviewed through full-text reading. A full article reading was required when the abstract mentioned the use of wearable sensors to monitor surgeon's exposure to risk factors associated with WRMSD, without explicitly excluding lower limb-related content.

3.2.3 Information Elicitation

Relevant information was systematically elicited from each selected paper in alignment with the objectives of the review. The collected information encompassed the following: publication year and country, type and body location of wearable sensors used on the surgeon's lower limbs, analyzed risk factors and ergonomic criteria for the analysis of the data related to lower limbs, surgery type, number of participants, task assessed, and the outcomes derived from the analysis of lower limb sensor's data.

3.3 RESULTS

This section presents the findings from the systematic review. The first subsection outlines the search results, which include the PRISMA flow diagram. The subsequent subsections focus on thematic content screening used to analyze and synthesize the elicited information. The thematic structure included the subsequent subsections:

- Descriptive statistical analysis of the publication year and country distribution.
- Descriptive analysis of the various wearable sensors used to collect data on surgeon's lower limbs. The elicited information is summarized in Table 3.1, which includes the bibliographic references, as well as details on the types of sensors used and their placement for lower limb data collection.
- Descriptive analysis of the risk factors and ergonomic criteria applied in evaluating the data recorded from the wearable sensors placed on surgeon's lower limbs.
- Descriptive analysis of the tasks assessed during the ergonomic risk assessment. The information elicited from the articles is summarized in Table 3.2, which presents the bibliographic references, types of surgeries assessed, number of participants, risk factors analyzed using data from wearable sensors on surgeon's lower limbs, the ergonomic assessment methods employed for evaluating the lower limbs, and the outcomes obtained from the analysis of the sensor data.

TABLE 3.1
Overview of the Sensors Used in the Lower Limbs

Study	Type of Sensor	Sensor's Placement
(Arrighi-Allisan et al., 2022)	IMU	Thighs, legs, and feet.
(Baird et al., 2021)	sEMG	Left and right biceps femoris.
(Gonzalez-Sanchez et al., 2017)	sEMG	Tibialis anterior, medial gastrocnemius, vastus medialis, and biceps femoris on the dominant side.
(Maurer-Grubinger et al., 2021)	IMU	Pelvis, thighs, legs, and feet.
(Pérez-Salazar et al., 2024)	EDA	Right ankle.

TABLE 3.2
Overview of the Studies Using Wearable Sensors for Lower Limb Ergonomic Risk Assessment

Study	Surgery Type	No. of Subjects	Risk Factor	Ergonomic Criteria	Output
(Arrighi-Allisan et al., 2022)	Functional endoscopic sinus surgery	6	Posture	REBA	Surgeons are exposed to different risk factors depending on their working posture. Seated positions tend to increase musculoskeletal strain in the upper limbs, whereas prolonged standing tends to contribute to strain in the lower limbs.
(Baird et al., 2021)	Laryngoscopy surgery	8	Muscle exertion	-	The left biceps femoris muscle showed a significant increase in the average sEMG RMS magnitude when the surgeon maintained bad posture during surgery, compared to when adopting a good posture.
(Gonzalez-Sanchez et al., 2017)	Robotic and laparoscopic surgery	1	Muscle exertion	-	Significant differences in mean muscle activation frequency were observed between the chief and assistant roles only during the robotic protocol. However, none of the lower limb muscles exceeded 25% MVC, indicating that muscle activation did not reach the minimum level of activation to cause muscle fatigue.
(Maurer-Grubinger et al., 2021)	Surgical dental treatment	15	Posture	Modified RULA	Although lower limb kinematics were recorded, the "leg score" in RULA remained fixed at +1 since the dental surgeons were seated throughout the procedure.
(Pérez-Salazar et al., 2024)	Robotic and conventional laparoscopic surgery	6	Stress and muscle exertion	-	EDA was significantly higher during laparoscopic procedures, particularly among novice surgeons, compared to robotic-assisted surgeries. In contrast, robotic surgery required greater muscle activity from the right vastus lateralis than conventional laparoscopy. Furthermore, complex and prolonged laparoscopic procedures resulted in increased loading of the gastrocnemius medialis muscle compared to robotic-assisted surgery.

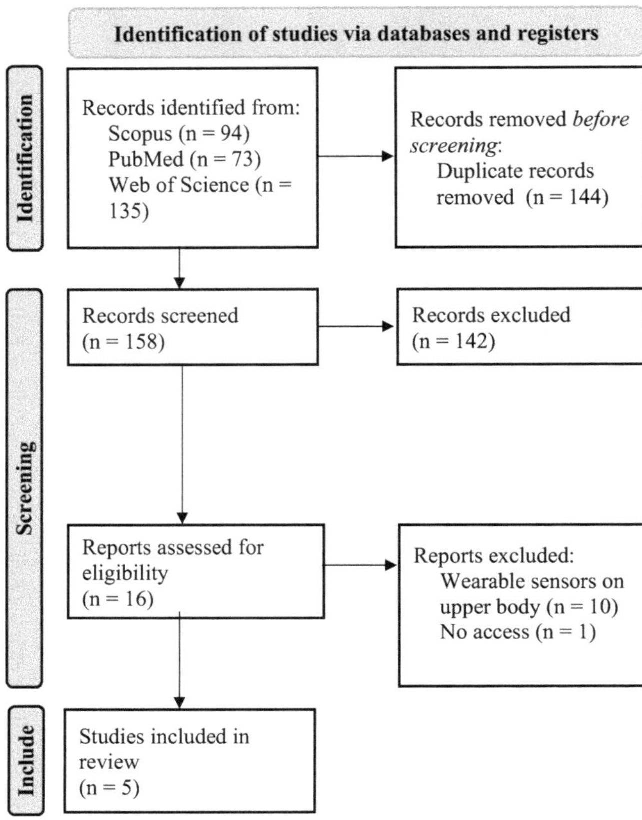

FIGURE 3.1 PRISMA flow diagram.

3.3.1 Search Results

The literature search across the three databases yielded 94 publications from Scopus, 73 from PubMed, and 135 from Web of Science. After removing duplicates, the total number of publications was reduced to 158. Then, the initial screening of titles and abstracts narrowed this down to 16 articles for more in-depth analysis. Following a full-text review, five articles were ultimately selected for inclusion in this systematic review. Figure 3.1 illustrates the PRISMA scheme summarizing the search results.

3.3.2 Year and Country of Publication

Figure 3.2 illustrates the distribution over time of studies included in this systematic review. The earliest study that met the inclusion criteria was published in 2017. The others were published in the last four years. This evolution suggests an increasing focus on the topic, underscoring the significance and timeliness of this work.

Figure 3.3 illustrates the geographical distribution of the studies. Spain contributed two studies, including the most recent publication from 2024. The USA also produced

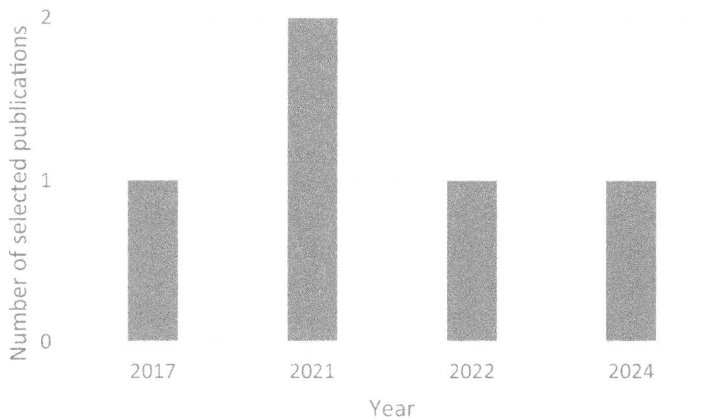

FIGURE 3.2 Number of selected publications per year.

FIGURE 3.3 Number of selected publications per country.

two studies, with publication dates ranging from 2021 to 2022. Additionally, Germany contributed one study in 2021. The geographical distribution of studies highlights the contribution of multiple countries to the field, revealing the existence of research efforts in North America and Europe. The observed distribution underscores the need for further research to address existing gaps, reinforcing the importance of continued investigation in this field.

3.3.3 Types of Wearable Sensors

In the context of lower limb ergonomic risk assessment of surgeons, three types of wearable sensors have been employed across different studies. Specifically, surface electromyography (sEMG) sensors were utilized in three studies, with one of these also incorporating an electrodermal activity (EDA) sensor (Baird et al., 2021; Gonzalez-Sanchez et al., 2017; Pérez-Salazar et al., 2024). Additionally, two studies utilized inertial measurement unit (IMU) sensors (Arrighi-Allisan et al., 2022; Maurer-Grubinger et al., 2021). Table 3.1 provides a detailed overview of the sensor placement locations on the lower limbs in each of the studies.

sEMG sensors provide an objective and quantitative method to evaluate muscle activity, movement patterns, and local muscle fatigue (Baird et al., 2021). This process involves positioning surface electrodes on the skin over target muscles, selected for their functional relevance to the analyzed activity. A grounding electrode is also placed to ensure accurate signal acquisition.

The IMU sensors measure body-posture angles by combining data from electromechanical sensors, such as accelerometers, gyroscopes, and magnetometers, within each unit (Maurer-Grubinger et al., 2021). These sensors are affixed to particular body segments using Velcro straps or integrated into a dedicated suit. The IMU solutions employed in the studies reviewed were commercially available systems (Arrighi-Allisan et al., 2022; Maurer-Grubinger et al., 2021).

Lastly, EDA sensors measure the electrical properties of the skin, providing insights into physiological responses associated with stress. In the study conducted by Pérez-Salazar and colleagues, one EDA activity sensor was employed, which captures a single-channel EDA signal using two Ag/AgCl electrodes (Pérez-Salazar et al., 2024).

3.3.4 Analyzed Risk Factors and Ergonomic Criteria

All the papers detail ergonomic risk assessments that quantify surgeon's exposure to risk factors, namely posture, muscle effort, and stress.

For assessing postural risk, the studies utilized IMU sensors to measure joint angles, followed by a comprehensive ergonomic risk assessment. When evaluating parameters associated with posture risk factor, one study applied a modified version of the Rapid Upper Limb Assessment (RULA), while another study applied the Rapid Entire Body Assessment (REBA) method (Arrighi-Allisan et al., 2022; Maurer-Grubinger et al., 2021). Both ergonomic risk assessments assign scores to specific body regions based on established thresholds for joint angles, enabling the correlation of observed body postures with associated risk levels. In the study conducted by Maurer-Grubinger and colleagues, the modified RULA provided a more detailed and sensitive analysis of joint angles by incorporating five distinct levels of investigation (Maurer-Grubinger et al., 2021). This approach allowed the researchers to transition from a high-level overview to a deeper exploration of parameters associated with posture risk factors, enabling the identification of subtle variations in musculoskeletal strain. The additional levels facilitated comprehensive comparisons between different postural conditions, yielding both a general summary through the RULA score and a more refined analysis using the "RULA steps" score. This enhanced method helped distinguish even minor differences in posture across various tasks.

For analyzing sEMG sensor data, the three studies took different approaches. Baird and colleagues calculated the average root mean squared (RMS) of the sEMG signal over five time periods, comparing these averages to the results of subjective pain scores to investigate the relationship between muscle activity and perceived discomfort during surgery (Baird et al., 2021). In the study conducted by Gonzalez-Sanchez and colleagues, the sEMG data was normalized to each participant's maximum voluntary contraction (MVC), reflecting the maximum electrical activity of a muscle (Gonzalez-Sanchez et al., 2017). The MVC for each muscle was obtained using Daniels' test, and two metrics were analyzed: median muscle activation frequency and median muscle activation slope. The median frequency provided an overview of the muscle activation pattern throughout the surgical procedure, while the slope, derived through linear regression, reflected the level of muscle fatigue. This method provided detailed information on muscle workload and fatigue development during different surgical protocols (Gonzalez-Sanchez et al., 2017). The study conducted by Pérez-Salazar and colleagues normalized sEMG data to participant's MVC and investigated localized muscle fatigue by analyzing changes in amplitude and median frequency across surgical procedures (Pérez-Salazar et al., 2024). The authors applied the Joint Analysis of Spectrum and Amplitude (JASA) method to quantify muscle fatigue, delivering a comprehensive view on muscle performance during surgical procedures (Dufaug et al., 2020).

For stress assessment, the study conducted by Pérez-Salazar and colleagues utilized EDA sensors to measure stress levels in surgeons during various tasks and procedures (Pérez-Salazar et al., 2024). Mean EDA values were calculated for each task and correlated with subjective workload ratings obtained through the SURG-TLX survey (Pérez-Salazar et al., 2024; Wilson et al., 2011). This correlation between objective stress measures and self-reported workload ratings provided a multidimensional understanding of the cognitive strain experienced by surgeons during surgical procedures.

3.3.5 Study Characteristics

The studies covered a variety of surgical types, including functional endoscopic sinus surgery, awake laryngeal surgery, surgical dental procedures, robotic surgery, and laparoscopic surgery. These studies were conducted in both real-world surgical settings and simulated environments. Table 3.2 offers an overview of these studies, specifically highlighting information related to the use of wearable sensors for collecting data on surgeons' lower limbs within the context of ergonomic risk assessment.

Two studies were conducted in real work environments. The study by Arrighi-Allisan and colleagues collected data using IMU sensors to perform an ergonomic risk assessment of the postures adopted by trainees and attendings during functional endoscopic sinus surgery (Arrighi-Allisan et al., 2022). The study conducted by Gonzalez-Sanchez and colleagues focused on recording muscle activity from four lower limb muscles of the dominant leg during robotic and laparoscopic surgeries

(Gonzalez-Sanchez et al., 2017). In this study, the surgeon undertook distinct roles in the two surgical types: as chief surgeon and assistant surgeon (Gonzalez-Sanchez et al., 2017). During robotic surgery, the chief surgeon operated the control device, while the assistant surgeon manipulated the robot's arms. In laparoscopic surgery, the chief surgeon performed the procedure, while the assistant held the camera.

In contrast, three studies were conducted in experimental tasks (Baird et al., 2021; Maurer-Grubinger et al., 2021; Pérez-Salazar et al., 2024). The study by Bair and colleagues recorded muscle activity from biceps femoris during a 10-minute simulated awake laryngeal surgery, comparing the effects of "good" versus "bad" posture on muscle activity (Baird et al., 2021). "Good posture" was defined as the surgeon standing adjacent to and directly in front of the examination chair, with the patient's head positioned at shoulder height. In contrast, "bad posture" involved the surgeon standing six inches laterally and away from the chair, with the patient's head positioned six inches below shoulder height (Baird et al., 2021). The study conducted by Maurer-Grubinger and colleagues recorded motion data for ergonomic risk assessment associated with the postures of oral and maxillofacial surgeons during standardized surgical tasks. These tasks were performed under two different dental treatment concepts: one in which the functional areas of the surgeon and assistant were divided, and another in which both shared the same workspace (Maurer-Grubinger et al., 2021).

Lastly, the study conducted by Pérez-Salazar and colleagues collected muscle activity and EDA data from surgeons during both simulator tasks and experimental model surgical procedures (Pérez-Salazar et al., 2024). The surgeons performed tasks in both laparoscopic and robotic-assisted surgeries. In the simulation phase, surgeons with varying levels of laparoscopic experience, but no prior exposure to robotic-assisted surgery, were assessed. In the experimental phase, surgeons executed procedures aligned with their surgical specialties. To ensure consistency, the operating table, monitor, screen, and console heights were adjusted according to each surgeon's anthropometrics.

3.4 DISCUSSION

The present systematic review aims to synthesize the evidence on the use of wearable sensors for lower limb ergonomic risk assessment of surgeons performing surgical procedures. Five studies were included in the review, highlighting the potential of wearable sensor technology to collect objective and feasible data for lower limb ergonomic risk assessments within surgeons. Despite the small number of studies, the increase of research published in the last four years highlights the increasing relevance of this area and validates the timeliness of this review.

The findings identified three types of wearable sensors to monitor surgeon's lower limbs: sEMG, IMU, and EDA sensors. Each sensor type provides distinct advantages in monitoring parameters related to surgeon's exposure to risk factors such as muscle exertion, posture, and stress.

IMU sensors, in particular, have emerged as optimal tools to apply in ergonomic risk assessments due to their integration of multiple types of sensors (e.g., accelerometers, gyroscopes, and magnetometers). This multi-sensor approach compensates for the limitations inherent in any single sensor type (Lim & D'Souza, 2020; Schall et al., 2016). The studies utilizing IMU devices have employed the RULA or REBA methods to evaluate parameters associated with posture risk factors (Arrighi-Allisan et al., 2022; Maurer-Grubinger et al., 2021). By recording work processes, these studies provided detailed insights into the percentage of time surgeons spent in different risk categories, enabling a more precise assessment of risk factor's exposure than could be achieved through observation alone. However, the findings underscore the need for the development of ergonomic risk assessment tools tailored to evaluate exposure to risk factors that contribute to the development of lower limb WRMSD. For instance, RULA is primarily designed for the upper limbs assessments, and its applicability to the lower limbs is limited; therefore, this tool may not adequately capture the physical strain placed on the surgeon's whole body during surgical procedures (Baird et al., 2021).

In addition to posture assessment, sEMG sensors were utilized to evaluate exposure to muscle exertion risk factor. The studies that included sEMG data revealed valuable insights into muscle activation patterns and muscle fatigue (Baird et al., 2021; Gonzalez-Sanchez et al., 2017; Pérez-Salazar et al., 2024). Notably, the study by Baird and colleagues found a lack of correlation between subjective reports of discomfort and sEMG RMS magnitude, suggesting that surgeons may endure microtraumas or muscle strain before consciously recognizing discomfort. This finding underscores the importance of objective monitoring of muscle activity in preventing WRMSD (Baird et al., 2021).

The assessment of stress levels during surgery was also explored using EDA sensors (Pérez-Salazar et al., 2024). High stress levels can negatively affect surgeon performance, potentially leading to errors, increased operative times, and compromised patient safety (Menke et al., 2024; Sujka et al., 2024). Given the precision required in many surgical procedures, effective stress management is essential, as it directly impacts fine motor skills and the surgeon's ability to execute delicate movements. Prolonged stress may also contribute to burnout, ultimately shortening a surgeon's career. Therefore, wearable sensors that continuously monitor stress indicators may significantly contribute in enhancing surgical performance and overall surgeon well-being (Pérez-Salazar et al., 2024).

In general, wearable sensors facilitate the continuous collection of procedural data instead of relying on intermittent snapshots, thereby providing a more accurate basis for ergonomic risk assessments (Asadi et al., 2021; Davila et al., 2021; Santos, Gabriel, Quaresma, & Nunes, 2024b). Importantly, these sensors facilitate the recording of lower limb postures without visual obstructions (Bartnicka et al., 2015; Pérez-Salazar et al., 2024; Santos, Gabriel, Quaresma, & Nunes, 2024b). Furthermore, the use of wearable sensors on lower limbs does not disrupt sterile scrubbing protocols during surgery, a limitation observed with wearable sensors on the upper limbs (Arrighi-Allisan et al., 2022; Davila et al., 2021; Yang et al., 2021).

Nevertheless, relying on wearable sensors requires addressing several technical challenges, including limited battery life, susceptibility to electromagnetic interference, connectivity issues, and potential inaccuracies due to skin movement or sensor displacement (Arrighi-Allisan et al., 2022; Asadi et al., 2021; Carbonaro et al., 2021). Moreover, participants' knowledge of the sensors may inadvertently influence their posture, as noted in the study by Arrighi-Allisan and colleagues, which evaluated surgeons during otolaryngologic surgical procedures (Arrighi-Allisan et al., 2022). Variability in sEMG sensor placement, along with differences in body mass index and overall muscularity, can affect the accuracy of sEMG data, as highlighted in the study conducted by Baird and colleagues (Baird et al., 2021).

Overall, the integration of wearable sensors addresses several limitations associated with self-reported or observational data collection methods for evaluating surgeon's working conditions. However, the small number of studies identified in this systematic review underscores the limited application of wearable sensors for lower limb ergonomic risk assessment of surgeons. This scarcity may reflect the broader underrepresentation of lower limb ergonomic risk assessments in the literature and the relatively recent advancements in wearable sensor technology. Additionally, compared to traditional ergonomic risk assessment methods, wearable sensors can be more complex and costly, which may hinder their widespread adoption at this stage.

Lastly, while wearable sensors represent a promising tool for assessing lower limb ergonomic risks in surgeons, further research is needed to establish standardized, lower-limb-specific ergonomic risk assessment protocols and to expand the application of this technology. Increased adoption of wearable sensors, combined with targeted ergonomic interventions, could play a critical role in reducing the prevalence of lower limb WRMSD and improving the long-term health and performance of surgeons.

3.5 CONCLUSION

The outcomes of this systematic review synthesize the evidence on employing wearable sensors for ergonomic risk assessment of surgeon's lower limbs. The analysis of five studies identified the various types of wearable sensors employed, providing an in-depth analysis of the methods applied. This information enhances understanding of the ongoing advancements in lower limb ergonomic risk assessment within surgical settings.

Integrating wearable sensors into surgical workflows represents a promising advancement in addressing lower limb ergonomic risk assessments and preventing lower limb WRMSD among surgeons. These technologies allow surgeons to make informed decisions during procedures by providing real-time monitoring of their exposure to risk factors. Furthermore, the data retrieved from wearable sensors can guide the development of specific lower limb ergonomic interventions, enhancing surgical efficiency while fostering a healthier workforce in the surgical field.

However, this systematic review also underscores the limited research on employing wearable sensors for lower limb ergonomic risk assessment of surgeons, emphasizing the need for further research. Expanding research in this field is crucial to fully realize the benefits of wearable sensor technology and ensure its effective integration into surgical practice. Future studies should focus on refining sensor technologies, developing lower-limb-specific ergonomic risk assessment tools, and validating their efficacy in diverse surgical contexts. By advancing these efforts, the field can move toward enhanced surgical outcomes and contribute to the occupational safety and well-being of surgical professionals.

REFERENCES

Aitchison, L. P., Cui, C. K., Arnold, A., Nesbitt-Hawes, E., & Abbott, J. (2016). The ergonomics of laparoscopic surgery: A quantitative study of the time and motion of laparoscopic surgeons in live surgical environments. *Surgical Endoscopy*, *30*(11), 5068–5076. https://doi.org/10.1007/s00464-016-4855-4

AlQahtani, S. M., Alzahrani, M. M., & Harvey, E. J. (2016). Prevalence of musculoskeletal disorders among orthopedic trauma surgeons: An OTA survey. *Canadian Journal of Surgery*, *59*(1), 42–47. https://doi.org/10.1503/cjs.014415

Alqahtani, S. M., Alzahrani, M. M., & Tanzer, M. (2016). Adult Reconstructive Surgery: A High-Risk Profession for Work-Related Injuries. *Journal of Arthroplasty*, *31*(6), 1194–1198. https://doi.org/10.1016/j.arth.2015.12.025

Alzahrani, M. M., Alqahtani, S. M., Tanzer, M., & Hamdy, R. C. (2016). Musculoskeletal disorders among orthopedic pediatric surgeons: an overlooked entity. *Journal of Children's Orthopaedics*, *10*(5), 461–466. https://doi.org/10.1007/s11832-016-0767-z

Arrighi-Allisan, A. E., Garvey, K. L., Wong, A., Filip, P., Shah, J., Spock, T., Del Signore, A., Cosetti, M. K., Govindaraj, S., & Iloreta, A. M. (2022). Ergonomic analysis of functional endoscopic sinus surgery using novel inertial sensors. *Laryngoscope*, *132*(6), 1153–1159. https://doi.org/10.1002/lary.29796

Asadi, H., Monfared, S., Athanasiadis, D. I., Stefanidis, D., & Yu, D. (2021). Continuous, integrated sensors for predicting fatigue during non-repetitive work: Demonstration of technique in the operating room. *Ergonomics*, *64*(9), 1160–1173. https://doi.org/10.1080/00140139.2021.1909753

Awosan, K. J., Yikawe, S. S., Oche, O. M., & Oboirien, M. (2017). Prevalence, perception and correlates of low back pain among healthcare workers in tertiary health institutions in Sokoto, Nigeria. *Ghana Medical Journal*, *51*(4), 164–174.

Baird, B. J., Tynan, M. A., Tracy, L. F., Heaton, J. T., & Burns, J. A. (2021). Surgeon positioning during awake laryngeal surgery: An ergonomic analysis. *Laryngoscope*, *131*(12), 2752–2758. https://doi.org/10.1002/lary.29717

Bartnicka, J., Zietkiewicz, A. A., & Kowalski, G. (2015). *Ergonomic analysis of surgeries with the use of wireless body postures measurement system. Innowacje w zarządzaniu i inżynierii produkcji* [Innovations in Manufacturing Management and Engineering], *2*, 361–371.

Carbonaro, N., Mascherini, G., Bartolini, I., Ringressi, M. N., Taddei, A., Tognetti, A., & Vanello, N. (2021). A wearable sensor-based platform for surgeon posture monitoring: A tool to prevent musculoskeletal disorders. *International Journal of Environmental Research and Public Health*, *18*(7), 3734. https://doi.org/10.3390/ijerph18073734

Cheng, J. W., & Mitomo, H. (2017). The underlying factors of the perceived usefulness of using smart wearable devices for disaster applications. *Telematics and Informatics*, *34*(2), 528–539.

David, G. C. (2005). Ergonomic methods for assessing exposure to risk factors for work-related musculoskeletal disorders. *Occupational Medicine (Oxford, England)*, *55*(3), 190–199. https://doi.org/10.1093/OCCMED/KQI082

Davila, V. J., Meltzer, A. J., Fortune, E., Morrow, M. M. B., Lowndes, B. R., Linden, A. R., Hallbeck, M. S., & Money, S. R. (2021). Intraprocedural ergonomics of vascular surgeons. *Journal of Vascular Surgery*, *73*(1), 301–308. https://doi.org/10.1016/j.jvs.2020.04.523

De Kok, J., Vroonhof, P., Snijders, J., Roullis, G., Clarke, M., Peereboom, K., Dorst, P. van., & Isusi, I. (2019). Work-related musculoskeletal disorders: prevalence, costs and demographics in the EU. In *European Agency for Safety and Health at Work*. Publications Office of the European Union. https://doi.org/10.2802/66947

Dufaug, A., Barthod, C., Goujon, L., & Marechal, L. (2020). New joint analysis of electromyography spectrum and amplitude-based methods towards real-time muscular fatigue evaluation during a simulated surgical procedure: A pilot analysis on the statistical significance. *Medical Engineering & Physics*, *79*, 1–9. https://doi.org/10.1016/j.medengphy.2020.01.017

Epstein, S., Sparer, E. H., Tran, B. N., Ruan, Q. Z., Dennerlein, J. T., Singhal, D., & Lee, B. T. (2018). Prevalence of work-related musculoskeletal disorders among surgeons and interventionalists: A systematic review and meta-analysis. *JAMA Surgery*, *153*(2), Art. e174947. https://doi.org/10.1001/jamasurg.2017.4947

Gonzalez-Sanchez, M., Gonzalez-Poveda, I., Mera-Velasco, S., & Cuesta-Vargas, A. I. (2017). Comparison of fatigue accumulated during and after prolonged robotic and laparoscopic surgical methods: A cross-sectional study. *Surgical Endoscopy and Other Interventional Techniques*, *31*(3), 1119–1135. https://doi.org/10.1007/s00464-016-5076-6

Halim, I., & Omar, A. R. (2012). Development of prolonged standing strain index to quantify risk levels of standing jobs. *International Journal of Occupational Safety & Ergonomics*, *18*(1), 85–96. https://doi.org/10.1080/10803548.2012.11076917

Jeong, S. C., Kim, S.-H., Park, J. Y., & Choi, B. (2017). Domain-specific innovativeness and new product adoption: A case of wearable devices. *Telematics and Informatics*, *34*(5), 399–412.

Kitchenham, B., & Charters, S. (2007). *Guidelines for Performing Systematic Literature Reviews in Software Engineering, version 2.3, technical report EBSE-2007-01.*

Kitchenham, B., Pretorius, R., Budgen, D., Brereton, O. P., Turner, M., Niazi, M., & Linkman, S. (2010). Systematic literature reviews in software engineering: A tertiary study. *Information and Software Technology*, *52*(8), 792–805. https://doi.org/10.1016/j.infsof.2010.03.006

Koutromanos, G., & Kazakou, G. (2020). The use of smart wearables in primary and secondary education: A systematic review. *Themes in ELearning*, *13*, 33–53.

Lim, S., & D'Souza, C. (2020). A narrative review on contemporary and emerging uses of inertial sensing in occupational ergonomics. *International Journal of Industrial Ergonomics*, *76*, Art. 102937. https://doi.org/https://doi.org/10.1016/j.ergon.2020.102937

Maurer-Grubinger, C., Holzgreve, F., Fraeulin, L., Betz, W., Erbe, C., Brueggmann, D., Wanke, E. M., Nienhaus, A., Groneberg, D. A., & Ohlendorf, D. (2021). Combining ergonomic risk assessment (RULA) with inertial motion capture technology in dentistry – using the benefits from two worlds. *Sensors*, *21*(12). https://doi.org/10.3390/s21124077

Menke, V., Hansen, O., Schmidt, J., Dechantsreiter, G., Staib, L., Davliatov, M., Schilcher, F., Hübner, B., Bianco, F., Kastelan, Z., Kulis, T., Hudolin, T., Penezic, L., Zekulic, T., Andelic, J., Juric, I., Puda, I., Siaulys, R., Venckus, R., … Samalavicius, N. E. (2024). The stress for surgeons: Exploring stress entities with the robotic senhance surgical system. *Journal of Robotic Surgery*, *18*(1), 94. https://doi.org/10.1007/s11701-024-01853-6

Nunes, I. L. (2009). FAST ERGO-X – A tool for ergonomic auditing and work-related musculoskeletal disorders prevention. *Work*, *34*(2), 133–148. https://doi.org/10.3233/WOR-2009-0912

Page, M. J., Moher, D., Bossuyt, P. M., Boutron, I., Hoffmann, T. C., Mulrow, C. D., Shamseer, L., Tetzlaff, J. M., Akl, E. A., Brennan, S. E., Chou, R., Glanville, J., Grimshaw, J. M., Hróbjartsson, A., Lalu, M. M., Li, T., Loder, E. W., Mayo-Wilson, E., McDonald, S., … McKenzie, J. E. (2021). PRISMA 2020 explanation and elaboration: Updated guidance and exemplars for reporting systematic reviews. *BMJ*, *372*. https://doi.org/10.1136/BMJ.N160

Pérez-Salazar, M. J., Caballero, D., Sánchez-Margallo, J. A., & Sánchez-Margallo, F. M. (2024). Comparative study of ergonomics in conventional and robotic-assisted laparoscopic surgery. *Sensors*, *24*(12). https://doi.org/10.3390/s24123840

Santos, C., Gabriel, A. T., Quaresma, C., & Nunes, I. L. (2024a). The use of wearable sensors for ergonomic risk assessment of surgical procedures: A literature review. *Human Factors and Systems Interaction*, *154*. https://doi.org/10.54941/ahfe1005376

Santos, C., Gabriel, A. T., Quaresma, C., & Nunes, I. L. (2024b). Exploring the use of wearable sensors for assessing risk factors during orthopedic surgeries: A protocol for data collection. *MethodsX*, 102994. https://doi.org/https://doi.org/10.1016/j.mex.2024.102994

Santos, C., Gabriel, A. T., Quaresma, C., & Nunes, I. L. (2024c). Risk factors, symptoms, and prevalence of lower limb work-related musculoskeletal disorders among orthopedic surgeons. *International Journal of Industrial Ergonomics*, *101*, Art. 103597. https://doi.org/https://doi.org/10.1016/j.ergon.2024.103597

Schall, M. C., Fethke, N. B., Chen, H., Oyama, S., & Douphrate, D. I. (2016). Accuracy and repeatability of an inertial measurement unit system for field-based occupational studies. *Ergonomics*, *59*, 591–602. https://api.semanticscholar.org/CorpusID:22592734

Sujka, J., Ahmed, A., Kang, R., Grimsley, E. A., Weche, M., Janjua, H., Mi, Z., English, D., Martinez, C., Velanovich, V., Bennett, R. D., Docimo, S., Saad, A. R., DuCoin, C., & Kuo, P. C. (2024). Examining surgeon stress in robotic and laparoscopic surgery. *Journal of Robotic Surgery*, *18*(1), 82. https://doi.org/10.1007/s11701-024-01834-9

Swank, K. R., Furness, J. E., Baker, E., Gehrke, C. K., & Rohde, R. (2022). A survey of musculoskeletal disorders in the orthopaedic surgeon: Identifying injuries, exacerbating workplace factors, and treatment patterns in the orthopaedic community. *JAAOS Global Research & Reviews*, *6*(5), e20. https://doi.org/10.5435/JAAOSGlobal-D-20-00244

Wilson, M. R., Poolton, J. M., Malhotra, N., Ngo, K., Bright, E., & Masters, R. S. W. (2011). Development and validation of a surgical workload measure: The Surgery Task Load Index (SURG-TLX). *World Journal of Surgery*, *35*(9), 1214. https://doi.org/https://doi.org/10.1007/s00268-011-1141-4

Xiao, Y., & Watson, M. (2017). Guidance on conducting a systematic literature review. *Journal of Planning Education and Research*, *39*(1), 93–112. https://doi.org/10.1177/0739456X17723971

Xu, A. L., Covarrubias, O. G., Yakkanti, R. R., Sotsky, R. B., & Aiyer, A. A. (2023). The biomechanical burden of orthopaedic procedures and musculoskeletal injuries sustained by orthopaedic surgeons: A systematic review. *JBJS Reviews*, *11*(1), e22. https://doi.org/10.2106/JBJS.RVW.22.00202

Yang, L., Wang, T., Weidner, T. K., Madura, J. A., Morrow, M. M., & Hallbeck, M. S. (2021). Intraoperative musculoskeletal discomfort and risk for surgeons during open and laparoscopic surgery. *Surgical Endoscopy*, *35*(11), 6335–6343. https://doi.org/10.1007/s00464-020-08085-3

4 Exploring Visual Tensions between Form and Color in Complex Static Systems

Lorena Olmos Pineda and Jorge Gil Tejeda

4.1 INTRODUCTION

There is a vast wealth of material culture generated by human beings throughout their existence. However, a particularly relevant category is that of artistic objects. Within this typology of objects, several important characteristics can be outlined, such as their high compositional complexity and their role as a medium for communication, both individual and collective expression. The communication fostered by this type of object can transcend borders and impact other cultures and micro-cultures in various ways. Some of these objects are so influential that they are considered transcultural. Despite generating multiple interpretations, they are both appreciated and valued. Comparatively speaking, artistic works, such as those depicted on canvases, are currently categorized from a human interaction perspective as complex objects with limited fine motor interaction; from this standpoint, they acquire the quality of being static. From a technological perspective, they are also considered static objects, since most virtual products today enable a wide range of human interactions, from fine motor interaction to gross motor interaction (Tejeda & Pineda, 2022). The points described above encompass some of their characteristics and their importance to people. For this reason, it is important to study this typology of complex objects from their configuration as static systems, as it is one of the main factors influencing how people interact with them. Understanding how these objects affect multiple individuals across different cultures and contexts can provide valuable insights for the creation of new products, not only physical but also virtual, where object identification is crucial. One discipline that involves the analysis of components to improve the creation of new products, systems, and services, while taking into account various human dimensions such as perceptual, emotional, needs, and capabilities, among other factors, is ergonomics. This discipline entails conducting research to understand how users interact with various objects, systems, and services, identifying issues related to communication, usability, interaction, recognition, fatigue, and difficulties encountered. Based on this information, prototypes can be developed, and tests can be conducted with real users, allowing for adjustments to be made based on feedback to ensure that the application is intuitive, accessible, and useful within the specific

DOI: 10.1201/9781003672036-4

context for which it is designed. One of the challenges faced by researchers is precisely to identify the components that most significantly impact interaction processes, analyze them, and adapt them to human needs.

Studies related to human interaction with various systems are complex and can be approached from a wide range of dimensions: psychological, sociocultural, ergonomic, technological, linguistic, and anthropological, to name a few. Cognitive ergonomics focuses on how people perceive, process, and understand information across a broad spectrum of artifacts. The interesting aspect of this discipline is that it allows for an examination of the components that shape interaction processes and how they affect human interaction. In this context, it is essential to identify the most important components of an artwork, such as color and shape. What is particularly intriguing is understanding how both components interact with each other and influence not only people's attention but also which of them plays a more dominant role in image recognition. In this regard, relevant studies have been conducted on face and object recognition, where the main interest lies in observing how different areas of the brain function. Notably, significant studies suggest that object recognition involves multiple components, such as perceptual processing, access to semantic memory, and visual memory (Behrmann, Winocur, & Moscovitch, 1992). In the recognition of specific factors such as a face, these are perceived more as a global configuration rather than as isolated components (Bruce, Doyle, Dench, & Burton, 1991), which Galton (1879) and Tanaka et al. referred to as a holistic perception of the face (Tanaka & Farah, 1993). The recognition of complex objects can be linked to certain types of specialized neurons located in the inferior temporal cortex (Desimone, 1991). The role of attention and different types of neurons located in the superior temporal sulcus respond to different views, which in turn affects object recognition (Perrett et al., 1991). Studying the problem of facial recognition as a two-dimensional problem with principal components (Turk & Pentland, 1991) implies a fragmentation of the image, where specialized brain resources are mentioned, such as memory for facial recognition as well as memory for other types of objects (Yin, 1969). In this regard, Biederman proposed that the perceptual recognition of objects is a process in which the input image is segmented by the brain into simpler components, which he termed "geons," facilitating recognition according to this theory (Biederman, 1987). This description highlights the vast amount of relevant research aimed at understanding object recognition. However, most of the aforementioned studies focus on the functioning of the brain. The present study aims to contribute from a different perspective: considering the configuration of the object and the responses given by individuals based on this configuration as a result of brain processing. Based on this configuration, the brain discriminates between certain components and makes determinations that are executed by individuals.

The study of visual tensions in static systems is crucial for optimizing interaction with various products and systems, and for understanding particularities in human brain function. These tensions, which arise from the perceptual balance or imbalance between visual elements, directly influence how users perceive and interact with the object. In this sense, research by Palmer highlights the importance of the hierarchical organization of object representation (Palmer, 1977), while the number of features affects the object recognition response time, as Bradshaw and Wallace discovered in their study of faces (Bradshaw & Wallace, 1971). Additionally, studies using different techniques have found that the characteristics of an object can be perceived through fragments and holistically (Matthews, 1978; Smith & Nielsen, 1970). In this regard, it is interesting to observe how object fragmentation influences recognition within a given time frame.

The study of these components and their perception has been conducted using various techniques. Among the most commonly used are PET scans, magnetic resonance imaging, electroencephalograms, and more recently, computational models (Marr, 1976). Understanding these dynamics of fragmenting complex systems through natural and artificial processes can enable various sectors to create artificially generated visual environments that are safe and ensure effective object recognition by the user. This is particularly important in environments where accurate identification, precision, reduced interference, and recognition time are crucial. Furthermore, knowledge of visual tensions helps anticipate potential usability issues, improving accessibility and the overall user experience.

The aim of this research is to explore, through cross-sectional studies, the visual tensions generated by the components of static systems considered transcultural and their influence on object recognition based on their artificial decomposition into basic shapes and color. The relevance of linking object identification to its decomposition into the components described above lies in understanding how people can be better aided in perceiving complex objects with which they interact in specific contexts, such as medical, military, transportation, or educational settings. This approach involves considering less common techniques that can contribute to the generation of bits of information.

4.2 VISUAL TENSIONS

The components of an image are diverse. However, visual processing allows for the discrimination of very similar characteristics within the image. Wertheimer (1938) observed apparent motion not of individual points, but of groups or "fields" in sequentially presented images (Wertheimer, 1938). Other studies have found that random dot textures, despite having similar power spectra, could still be effortlessly discriminated based on conspicuous local features such as quasi-linearity, edges, closure, and connectivity (Julesz, 1982). Additionally, there are studies related to information processing, which distinguish at least three important levels to consider from a human perspective, as well as in relation to a computational theory (Marr & Poggio, 1976). At the lowest level is the analysis of basic components and circuits, which describes how neurons function. The second level is the study of specific mechanisms: summators, multipliers, and memories, which are assemblies made from basic components (Marr, 1977).

Another theory suggests that the brain's first operation on an image is to transform it into a primitive but rich description of how intensities change across the visual field, capturing the key aspects necessary for image analysis. Marr referred to this as a "primal sketch" (Marr, 1976). The traditional approach in computer vision assumes that the essence of such decoding is a process called segmentation, whose purpose is to divide a primal sketch—or more generally, an image—into regions that are meaningful, perhaps as physical objects (Marr, 1982). Image segmentation techniques have therefore been used to study the perceptual process. In this context, a fundamental component in image segmentation is color and the perceptual variations it offers in image composition.

Von Helmholtz (2013) conducted studies on how the human eye perceives light and images by exploring the physiological and optical mechanisms that integrate visual perception, including how light interacts with the eye, how images are formed on the retina, and how the brain processes this information to produce visual experience (Von Helmholtz, 2013). For David Marr, there are two complementary approaches to brain theory. The "bottom-up" approach starts from detailed information about individual neurons and attempts to analyze their behavior when connected in simple or highly regular neural networks. The "top-down" approach employs knowledge from artificial intelligence (AI), attempting to analyze which pattern of interactions between what type of subsystems might be necessary to achieve a general process (Arbib, 1979).

To study visual tensions, various techniques can be employed. In this regard, perceptual training or visual decomposition and association is useful for helping individuals improve their visual identification skills through gradual exposure and recognition of patterns in controlled situations. Cognitive psychology utilizes this technique in research on visual perception and cognition to understand how the brain processes and recognizes patterns and objects.

4.3 CONSIDERATIONS IN HUMAN INTERACTION WITH STATIC SYSTEMS

One of the most important factors to consider in this study was related to the optimal times for image identification. A complex task that individuals perform in their daily lives is the natural processing of scenes. These are deemed complex because, in contemporary dynamics, people do not take the time to observe in detail the vast amount of stimuli present in their environment. Typically, they are perceived while engaging in everyday activities such as walking, driving, or running. In addition to the complexity of the images, there are components that influence their processing, such as exposure time, whether the image is stationary or in motion, and the number of components in the image, among others. To measure visual processing speed, techniques such as behavioral methods and neuroimaging can be employed (Thorpe & Fabre-Thorpe, 2002).

There are various relevant studies that have indicated that an exposure time of 20 ms is sufficient for the visual system to retain relevant information from an image (Thorpe, Fize, & Marlot, 1996). However, processing a complex natural image typically requires an average of 150–200 ms (Thorpe, Fize, & Marlot, 1996; Liu, Harris, & Kanwisher, 2002; Mouchetant-Rostaing, Giard, Delpuech, Echallier, & Pernier, 2000). To measure visual processing of an image, tasks are utilized that require individuals to make some type of decision regarding the stimulus (Kirchner & Thorpe,

2006). In this regard, not only should image processing be considered, but also reaction time. Reaction time has been extensively studied. Donders is recognized as a pioneer in measuring cognitive processing time and its corresponding motor response based on experimental data. His method involved decomposing the process into stimulus detection, decision-making, and execution (Prinz, 1997). The approach by De Lotze and James, on the other hand, focused on the study of voluntary actions (James, 1980).

In this context, studies have shown that individuals take between 450 and 250 ms (Kirchner & Thorpe, 2006). As can be seen, visual information processing and reaction are essential components. Thus, a time frame of 5 seconds was proposed as sufficient for a person to observe and reflect on the proposed complex static systems (CSSy).

4.4 METHODOLOGY

The study was divided into five phases. In the first phase, an empirical approach was employed in which volunteers self-reported statistical data such as ID, age, gender, and emotional response valence, among others. The context was the everyday physical space. In the second phase, the group of individuals was assigned a non-complex task: essentially, the exploration of 20 CSSy corresponding to the works of the artist van Gogh and proceeding to identify them based on the observation of segmented images derived from color and the contours defined by variations in the luminance of the original image. The CSSy were determined by artistic works with color ranges of 3,700–268,000 and shapes of 17–2,900. The segmentation of the image into colors and contours was determined using the *KMeans* algorithm and the *sklearn* library for colors; similarly, the number of contours was established using the OpenCV library and the *PIL* library. The group of images was classified by colors and contours to observe which image components most facilitated their identification. In the third phase, the 20 selected images as complex systems were presented to the participants with a fixed time of 5 seconds. In the fourth phase, participants were required to select from a set of two segmented images the one they recognized as closest to the one observed. The time to provide their response was also 5 seconds. In the fifth phase, the participants' responses were analyzed concerning their accuracy in identifying the images.

4.5 DESCRIPTION OF THE PROCESS

Fifteen images by the artist van Gogh were selected, which, due to their compositional qualities, fall within the category of CSSy. Subsequently, the images were classified into four categories based on their content: people, animals and plants, objects and landscapes.

In the first category, related to people in the CSSy (see Figure 4.1), the original image was segmented into lines determined by chromatic changes within the system (see Figure 4.1). This was done considering the theories described by Yin (1969), Biederman (1987), and other researchers, who describe the process of image input

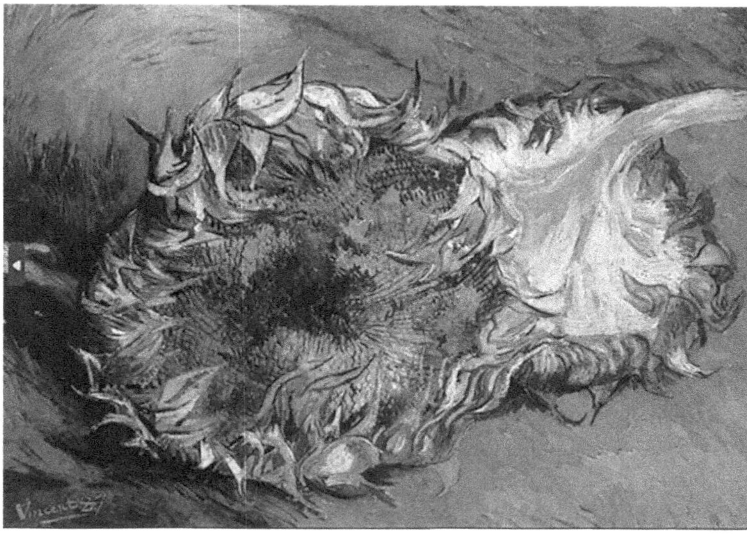

FIGURE 4.1 Vincent van Gogh. Title: *Two Cut Sunflowers*. August–September 1887, Paris Oil on canvas, 43 × 61 cm. Metropolitan Museum of Art, New York. ID: flower13; No. of Unique Color: 168755.

to the brain as being based on its decomposition into simpler forms. For this purpose, an approach based on contour segmentation and simplification was employed for the detection of contours in an image, using the *OpenCV* library. Volunteers were required to identify the real image to which the segmented line image belonged. The two options available for identification were two similar images, which depicted two people in similar positions, with both images presented at the same level of realistic representation as painted by the author (see Figure 4.1). The exposure time for the base image was 5 seconds, and the time given to the volunteers to identify it was also 5 seconds.

In the second category, corresponding to animals and plants, the original image was segmented based on a color mapping where the four dominant colors were considered. This was accomplished using the *KMeans* algorithm, the *PIL* library for image handling, and *sklearn* for color processing. Color mapping is a technique used in various fields, such as computer graphics, computer vision, and image processing, to transform the colors of an image or visual representation according to a predefined function or color scheme. In this sense, color mapping generated a segmented image. Additionally, for this process, the *KMeans* clustering algorithm was utilized, which traced the most representative colors of each image (see Figure 4.3).

Volunteers had to identify the image corresponding to the segmentation by lines determined by colors from the segmented color image (see Figures 4.1 and 4.3). In this regard, segmentations with line densities corresponding to the real one were chosen. The exposure time for the base image was 5 seconds, and the time given to the volunteers to identify it was also 5 seconds.

In the third category of objects and landscapes, the original image was segmented based on the dominant color of each image. The dominant color obtained occupied the same position in the original composition (Figure 4.2). Volunteers were required to identify the original image from two options: one was a result of line fragmentation of the original image (see Figure 4.1), and the other was a result of color mapping fragmentation of the original image (see Figure 4.3). The exposure time for the base

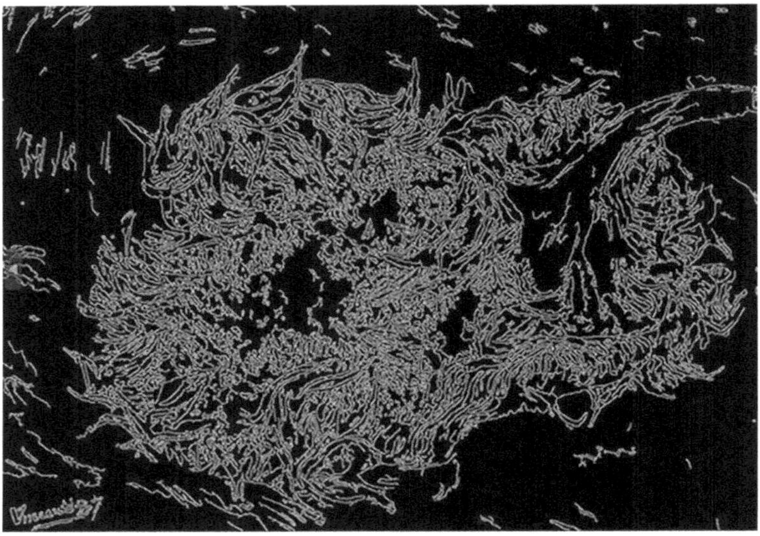

FIGURE 4.2 ID: flower13; No. of Figures: 1450. Main image of Category 1.

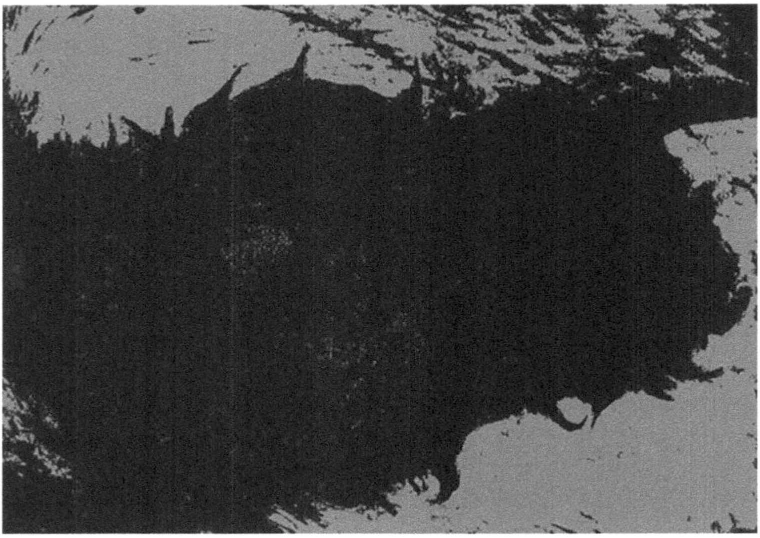

FIGURE 4.3 ID: flower13; Dominant color. KMeans algorithm. Main image of Category 3.

image was 5 seconds, and the time allocated for volunteers to identify it was also 5 seconds.

For counting unique colors and geometric shapes, the images were converted to the RGB color space, and *np.unique* was used to count the distinct colors as a pixel array. A pixel array is a data structure in which each pixel of an image is represented as a value or set of values. In the context of image processing with Python, pixel arrays are commonly used for manipulating and analyzing images. For the counting of geometric shapes, the contours of the image were first detected to locate the shapes: triangles, squares, rectangles, and circles.

Based on the researchers described earlier, it was established that volunteers had to utilize resources such as memory, processing of the input image through the visual system, and spatial localization, and the perceiver had no prior experience with the exercise, thus lacking habituation to the stimulus.

The total number of volunteers comprised 56.521% women, 39.13% men, and 4.347% from the LGBTQAI+ community. It was observed that 65.797% of users correctly identified the images across the three categories, indicating a highly sophisticated level of the visual system for processing information. It is also observed that 5 seconds is a generous amount of time to observe the complex qualities of the CSSy (image), which aids in its identification, at least in situations where one must identify it from a secondary image derived from the original input image. It is important to consider that for the execution of the activity, there was no habituation to the image, meaning there was no memorization of either the input image or the images derived from the original. The category with the highest accuracy was Category 3, which achieved 100% accuracy. This was followed by Category 2, with an accuracy of 70.434%, and finally, Category 1, with an accuracy of 55.652%. Notably, in Category 3, both responses were correct, as both images presented as possible choices were products of the fragmentation of the original image—one derived from lines based on variations in color, and the other based on a color mapping of the four dominant tones of the image. In this regard, 71.304% of volunteers selected the image generated from the color mapping, suggesting that in time-constrained situations where the image behaves as a CSSy, color may be a factor that individuals tend to prioritize for recognition, compared to the 6.086% of volunteers who chose the image generated from lines. Furthermore, 22.608% of volunteers were unable to complete their response due to running out of time. Among those who chose color as the key component for image recognition, 43.478% were women, 39.130% were men, and 4.347% identified as part of the LGBTQIA+ community. In relation to the group that chose lines as the key component for image recognition, 13.043% were women, 8.695% were men, and 0% identified as part of the LGBTQIA+ community. Additionally, it can be observed that in all four categories—people, animals and plants, objects, and landscape—the dominant component is color, as shown in Figures 4.2, 4.3, 4.4, and 4.5. In this context, a unique color refers to each of the shades detected by the algorithm in the composition, which are, in strict terms, distinct (see Table 4.1).

It was also observed that in both Categories 1 and 2, which had correct responses, the accuracy did not reach 100%. This indicates that there was always a choice of an incorrect response by the perceivers. In this regard, in Category 1, 11.304% of the

TABLE 4.1
Shows the Components: No. of Unique Colors; No. of Triangles; No. of Squares; No. of Rectangles; No. of Circles; No. of Figures

TYPE	ID IMAGE	No. of unique color	No. of triangles	No. of squares	No. of rectangle	No. of circles	No. Figs.	Dom. Color 1 RGB Code	Dom Color 2	Dom. Color 3	Dom Color 4
Person	auvers24	218166	33	6	222	2065	2326	177 167 139	76 71 64	121 122 102	24 24 31
Person	copies01	183358	12	6	101	2029	2148	41 37 32	148 140 103	96 86 71	222 201 139
Person	copies13	251097	23	12	122	2035	2192	189 151 96	136 176 192	46 38 32	121 101 81
Person	Paris10	225186	7	4	63	1287	1361	99 92 70	149 131 99	43 42 39	186 180 144
Person	selfpo18	132826	1	1	21	543	566	125 95 55	237 188 88	76 59 53	165 141 82
Animals	2hague09	3774	0	0	1	16	17	114 106 91	203 195 174	55 48 43	160 152 132
Animals	3arles26	107962	6	2	52	573	633	123 82 54	195 209 185	182 132 76	76 51 38
Animals	arles65	237378	30	11	123	1868	2032	202 204 183	109 84 59	37 30 22	173 151 112
Animals	auvers26	154749	6	4	70	959	1039	110 66 45	161 133 89	49 36 27	188 170 133
Animals	Flower13	168755	17	5	66	1362	1450	83 111 152	33 32 32	187 154 93	91 85 74
Objects	1arles12	175794	10	4	62	747	823	25 25 31	194 168 88	188 188 183	103 94 95

Objects	1remy10	26569	12	4	63	324	403	233 225 199	39 34 26	196 187 162	134 126 108
Objects	3arles29	124215	33	8	107	1529	1677	173 144 128	119 89 73	36 25 22	221 196 178
Objects	4earlie1	256808	8	3	119	889	1019	127 102 63	118 136 172	67 51 32	179 169 111
Objects	Arles59	122717	13	6	65	1363	144	97 94 69	132 140 102	179 194 154	59 44 30
Palsage	1arles07	268457	23	5	91	2231	2350	158 143 56	217 201 81	72 79 33	143 170 183
Palsage	1asylu01	212421	59	25	162	2662	2908	80 77 65	132 126 104	25 25 24	199 187 154
Palsage	1asylu07	198268	16	10	78	2226	2330	42 41 37	100 116 134	177 149 99	92 85 70
Palsage	1asylu09	203238	28	12	130	2766	2936	133 116 81	42 33 23	191 179 123	89 68 49
Palsage	auvers28	241683	31	16	140	2613	2800	204 196 185	124 124 98	78 80 52	164 160 144

Source: Data obtained with the *KMeans* algorithm.

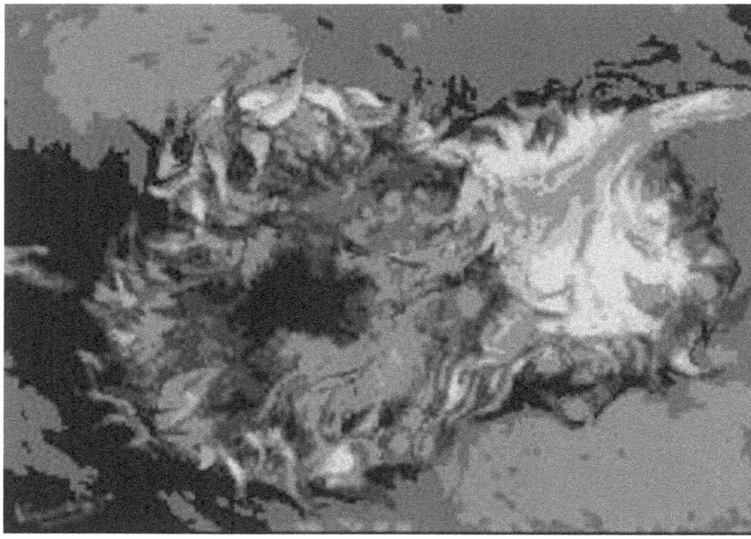

FIGURE 4.4 ID: flower13; Color mapping.

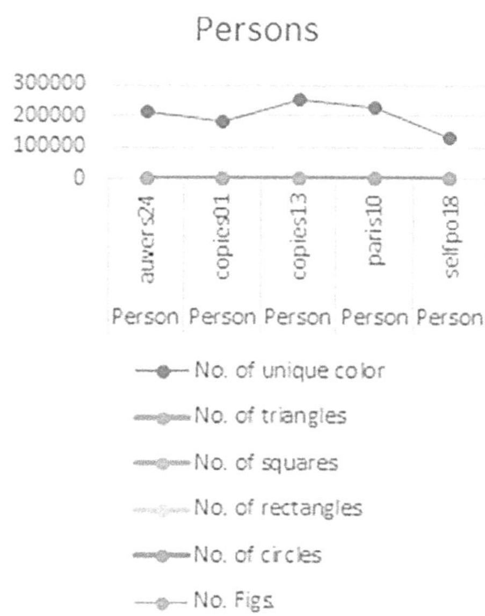

FIGURE 4.5 Displays the dominant component in the category: People.

total volunteers chose an incorrect response, of which 43.478% were women and 17.391% were men. In Category 2, 13.043% of the total volunteers chose an incorrect response, of which 21.739% were women and 26.086% were men. These results indicate that Category 2 was where the most errors occurred in image identification.

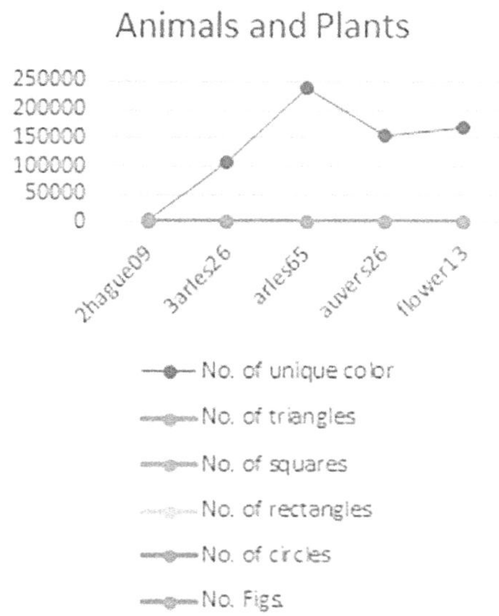

FIGURE 4.6 Shows the dominant component in the category: Animals and Plants.

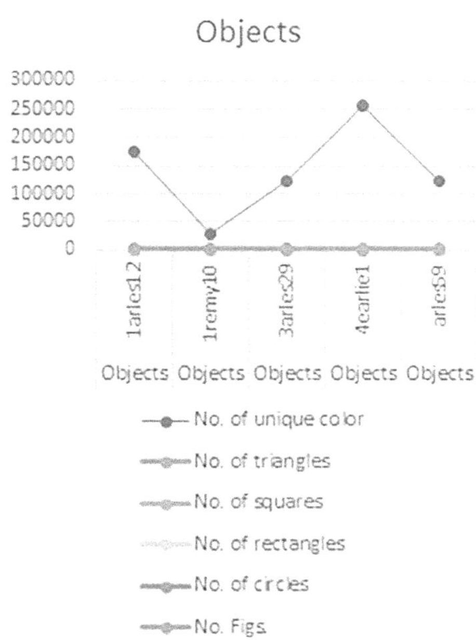

FIGURE 4.7 Displays the dominant component in the category: Objects.

This category featured animals and objects. In this category, the initial image was presented fragmented by color mapping, and subsequently, the real image had to be identified as fragmented by lines, resulting in a much more complex object. As shown in Figure 4.8, the fragmentation into lines based on variations in luminosity

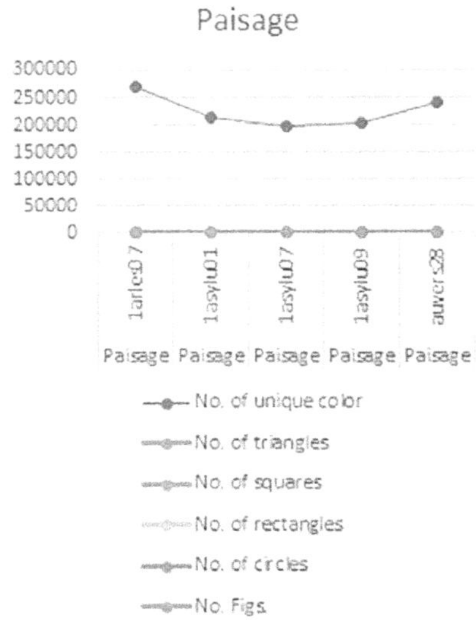

FIGURE 4.8 Displays the dominant component in the category: Landscape.

of the composition generated a CSSy of 2,148 components. This suggests that although the forms are familiar, fragmentation into lines and shapes complicates their identification.

It was also noted that the female participants had more difficulty identifying the original image from an image fragmented by lines (greater difficulty) and performed better when identifying the image from an image fragmented by color. In contrast, the male participants performed better in identifying the image from an image fragmented by lines compared to the image fragmented by color. Furthermore, it was observed that the process involved the observation of a CSSy and its identification with static systems of lesser complexity. However, this complexity does not stem from the number of components but from the visual processing that the brain performs on the image. The number of lines and shapes that make up (see Figure 4.8) has a lower number of components (see Table 4.1) compared to the number of unique colors that the same image could contain (see Figure 4.9), where 183,358 unique colors were identified (see Table 4.1).

In the case of Category 2, the observer's process involves observing a less CSSy before progressing to a more CSSy. In this sense, identification begins with the observation of the image fragmented by the dominant color (see Figure 4.2) and culminates in the identification of the more complex image, which is integrated through its fragmentation into lines and shapes (see Figure 4.1). This indicates that identifying the input static system is fundamental for proceeding to identify a more complex system. Conversely, this demonstrates a higher range of error, meaning incorrect identification of the primary image, which hinders the association with either a less complex or more complex image.

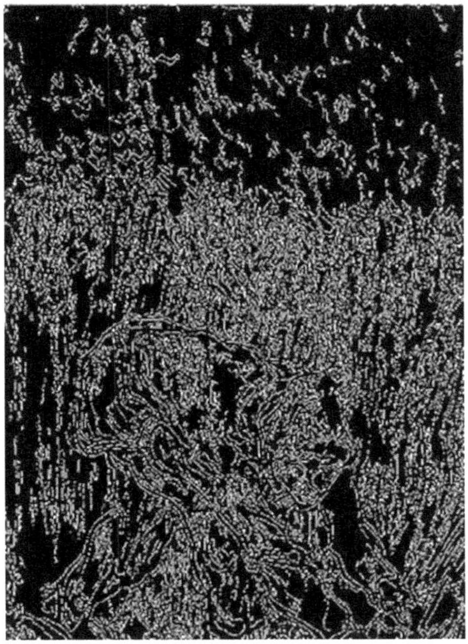

FIGURE 4.9 ID: Copies01. Main image of Category 1. Decomposition into lines using the OpenCV library.

FIGURE 4.10 First choice option.

FIGURE 4.11 Second option for selection.

Moreover, it was observed that Category 1 had the highest level of unanswered questions, at 33.043%, followed by Category 3 at 22.608%, and Category 2 at 16.521%. This may suggest greater difficulty in identifying the image within that category.

Additionally, the exercise presents a series of unconventional or general images, as it reflects the very particular perception of the artist van Gogh. This implies a recognition described by some researchers in the field of cognitive psychology, who propose that the categorization of objects does not always occur at a general or basic level; there are instances where much more specific or subordinate categories are perceived (Rosch, Mervis, Gray, Johnson, & Boyes-Braem, 1976). In this sense, the recognition performed by the volunteers occurred at a subordinate level.

From Biederman's perspective, who proposed that perceptual recognition of objects is a process where the input image is segmented (Biederman, 1987), it is observed that this does not always apply, at least when utilizing the technique of perceptual training. Throughout the exercise, it was noted that individuals, to reduce identification error, must first recognize the object, whether at a general-basic level or a particular-subordinate level. Here, the number of compositional elements does not determine identification, but rather the qualities of the compositional elements. It is observed that color, in the case of CSSy, aids in better image identification.

It is important to emphasize that we are discussing a very particular type of system, which comprises ranges of 20,000 unique colors and over 2,000 shapes. This information is fundamental for identifying complex objects in contexts where such identification is crucial. This was identified in Category 3 of images, where the input image was fragmented by the dominant color, requiring recognition to distinguish between two possible options. Both were correct, except that one was fragmented by lines and shapes, while the other was fragmented by color mapping. As a result, volunteers demonstrated a higher recognition of the image based on color mapping.

4.6 BEHAVIOR OF THE DYNAMIC IDENTIFICATION HUE

The results reveal an interesting pattern. The preference for color as a supernormal stimulus, as defined by Ramachandran and Hirstein (1999), serves as a component within the artistic domain that amplifies certain characteristics of the artwork and aids in capturing the viewer's attention. In this study, it was observed that, in the presence of CSSy, color acts as a key component or supernormal stimulus for image recognition, which may correlate with gender. Specifically, it was noted that women experienced greater difficulty in identifying the original image from a line-fragmented image (greater difficulty) and performed better in identifying the image from a color-fragmented image. In contrast, men exhibited the opposite identification behavior.

In this context, the behavior of dynamic hue identification can be conceived as the tensions generated during the identification process, wherein it is assumed that both shape and color are components that affect the perceiver. These perceptual tensions arise, impacting the identification of an input image, where a higher percentage of volunteers prioritized color over shape under conditions of CSSy within a controlled timeframe. The number of compositional elements does not determine image identification; rather, it is the qualities of the compositional elements that matter. To reduce the error of identification, the object must first be recognized at either a general-basic level or a specific-subordinate level. Consequently, the interaction of these components led to the finding that a lower identification of a CSSy as the input image increases the identification error in a derived image from that system. Conversely, a higher identification of a CSSy as the input image decreases the identification error in a derived image from that system. In this regard, the information recorded indicates that the complexity in this type of system does not stem from the number of components. Rather, it lies in the manner in which the brain performs visual processing, with color being the component that most facilitated the identification of the correct image. The idea that the tensions generated during the identification process are influenced by shape and color is relevant. This may suggest that the brain is not only collecting visual information but also engaged in a constant selective processing among different visual components. In this context, prior familiarity and recognition may facilitate more accurate identification processes, reinforcing the idea that the brain is a highly specialized mechanism for resource management.

REFERENCES

Arbib, M. (1979). Feature detectors, visumotor coordination and efferent control. In *Recognition of pattern and form: Proceedings of a conference held at the University of Texas at Austin* (pp. 100–110). Springer Berlin Heidelberg.

Behrmann, M., Winocur, G., & Moscovitch, M. (1992). Dissociation between mental imagery and object recognition in a brain-damaged patient. *Nature*, *359*(6396), 636–637.

Biederman, I. (1987). Recognition-by-components: A theory of human image understanding. *Psychological Review*, *92*(2), 115.

Bradshaw, J., & Wallace, G. (1971). Models for the processing and identification of faces. *Perception and Psychophysics*, *9*, 443–448.

Bruce, V., Doyle, T., Dench, N., & Burton, M. (1991). Remembering facial configurations. *Cognition*, *38*(2), 109–144.

Desimone, R. (1991). Face-selective cells in the temporal cortex of monkeys. *Journal of Cognitive Neuroscience*, *3*(1), 1–8.

Galton, F. (1879). Composite portraits, made by combining those of many different persons into a single, resultant figure. *Journal of the Anthropological Institute*, *8*, 132–144.

James, W. (1980). *Principles of psychology*. Henry Holt.

Julesz, B. (1982). The role of terminators in preattentive perception of line textures. In D. G. Albrecht (Ed.), *Recognition of pattern and form (Lecture notes in biomathematics*, Vol. 44). Springer Berlin Heidelberg.

Kirchner, H., & Thorpe, S. (2006). Ultra-rapid object detection with saccadic eye movements: Visual processing speed revisited. *Vision Research*, *46*(11), 1762–1776.

Liu, J., Harris, A., & Kanwisher, N. (2002). Stages of processing in face perception: An MEG study. *Nature Neuroscience*, *5*(9), 910–916.

Marr, D. (1976). Early processing of visual information. *Philosophical Transactions of the Royal Society of London. Biological Sciences*, *942*, 483–519.

Marr, D. (1977). Artificial Intelligence—A personal view. *Artificial Intelligence*, *9*, 37–48.

Marr, D. (1982). Visual information processing: The structure and creation of visual representations. In D. G. Albrecht (Ed.), *Recognition of pattern and form (Lecture notes in biomathematics, Vol.* 44). . Springer Berlin Heidelberg. https://doi.org/10.1007/978-3-642-93199-4_4

Marr, D., & Poggio, T. (1976). From understanding computation to understanding neural circuitry. *Neurosciences Research Program Bulletin*, *15*, 470–488.

Matthews, M. (1978). Discrimination of Identikit construction of faces: Evidence for a dual processing strategy. *Perception and Psychophysics*, *23*, 15–161.

Mouchetant-Rostaing, Y., Giard, M., Delpuech, C., Echallier, J., & Pernier, J. (2000). Early signs of visual categorization for biological and non-biological stimuli in humans. *Neuroreport*, *11*(11), 2521–2525.

Palmer, S. (1977). Hierarchical structure in perceptual representation. *Cognitive Psychology*, *9*(4), 441–474.

Perrett, D., Oram, M., Harries, M., Bevan, R., Hietanen, J., Benson, P., & Thomas, S. (1991). Viewer-centred and object-centred coding of heads in the macaque temporal cortex. *Experimental Brain Research*, *86*, 159–173.

Prinz, W. (1997). Why Donders has led us astray. *Advances in Psychology*, *118*, 247–267.

Ramachandran, V., & Hirstein, W. (1999). The science of art: A neurological theory of aesthetic experience. *Journal of Consciousness Studies*, *6*(7), 15–51.

Rosch, E., Mervis, C., Gray, W., Johnson, D., & Boyes-Braem, P. (1976). Basic objects in natural categories. *Cognitive Psychology*, *8*, 382–439.

Smith, E., & Nielsen, G. (1970). Representations and retrieval processes in short-term memory: Recognition and recall of faces. *Journal of Experimental Psychology*, *85*, 397–405.

Tanaka, J., & Farah, M. (1993). Parts and wholes in face recognition. *The Quarterly Journal of Experimental Psychology Section A*, *46*(2), 225–245.

Tejeda, J., & Pineda, L. (2022). User experiences derived from mass-distribution virtual products are integrated by the spatio-temporal component. *Usability and User Experience*, *39*, 119–125.

Thorpe, S., & Fabre-Thorpe, M. (2002). Fast visual processing and its implications. In M. Arbib (Ed.), *The handbook of brain theory and neural networks* (pp. 441–444). MIT Press.

Thorpe, S., Fize, D., & Marlot, C. (1996). Speed of processing in the human visual system. *Nature*, *381*(6582), 520–522.
Turk, M., & Pentland, A. (1991). Eigenfaces for recognition. *Journal of Cognitive Neuroscience*, *31*(1), 71–86.
Von Helmholtz, H. (2013). *Treatise on physiological optics* (Vol. III). Courier Corporation.
Wertheimer, M. (1938). *Laws of organization in perceptual forms.* Kegan Paul, Trench, Trubner & Company.
Yin, R. (1969). Looking at upside-down faces. *Journal of Experimental Psychology*, *81*(1), 141.

5 The GALA Framework
Qualitative Expert Evaluation and First Extension

Daniele Ruscio, Marvin Schopp, Marie Oppitz, Daniel Dreyer and Benedikt Petermeier

Level of Automation (LoA) frameworks serve to represent the varying levels of assistance a human operator can receive from an automated system (Frohm et al., 2008). These frameworks are useful tools for designing automated systems, prompting reflection on the appropriate LoA for specific tasks (Save et al., 2012). Sheridan et al. (1978) are credited with creating what is often considered the first LoA framework (Vagia et al., 2016). Their model introduced ten levels ranging from "Human must take all decisions and actions" to "Computer decides everything", clearly defining authority at each stage, but focusing only on a specific type of automation that considers both decision selection and action implementation (Vagia et al., 2016). Parasuraman et al. (2000) expanded the model to include four types of automation: information acquisition, information analysis, decision selection, and action implementation. These information processing stages do not always occur in a strict sequence, and for each type, an automation level can be assigned on a continuum ranging from "high" to "low" (Parasuraman et al., 2000).

These foundational models have inspired further research and additional LoA frameworks. For instance, the Level of Automation Taxonomy (LOAT) by Save et al. (2012) offers a comprehensive two-dimensional taxonomy to classify and compare different kinds of technologies in order to support the decision about the extent of automation support. Endsley and Kaber (1999) proposed a framework similar to Sheridan's, with ten levels ranging from "manual control" to "full automation", and with a second dimension focusing on the "roles" that the "Human/Computer" have in four functions: monitoring, generating, selecting, and implementing. Other models, like the PACT framework (Bonner et al., 2000; Taylor, 2002), focus on the concept of "authority/autonomy" of the operator and the computer, with three primary modes: commanded, assisted, and automatic.

In addition, since different fields have distinct needs that must be considered when designing and evaluating automated systems (Andersson, 2011), different frameworks were generated ad-hoc, sometimes making them applicable only in a specific domain. For instance, in the automotive industry, frameworks have been developed to categorize only the automation of driving tasks. The SAE framework, specifically designed for automated driving systems, addresses the unique conditions of automated driving (SAE International, 2021). Robotics is another field with very

 DOI: 10.1201/9781003672036-5

specialized LoA frameworks. While robots were originally used primarily in manufacturing, their roles have expanded (Janssen et al., 2019). Consequently, frameworks have been developed to assess human-robot interactions (Müller et al., 2017). In the medical field, surgical robots can facilitate operations and find increasing usage in various forms of surgery (Lee et al., 2024). The application of medical robots is safety critical and entails regulatory and legal concerns about automation levels that need to be addressed within that domain (Yang et al., 2017). Similarly, in military aviation, automated weapon systems are integrated, and the operators need to handle threat scenarios and not all the frameworks can apply to such situations (Boy, 1998). It is therefore not surprising that several LoA frameworks targeting specific technology were developed within a given domain (Lee et al., 2024).

Despite the variety of existing frameworks/taxonomies/models, and the extensive and significant merits of LoA concepts (Kaber, 2018), some common concerns and potential limitations about LoA concepts can still be identified:

1. Representing certain functions that automation can be applied to. Certain LoA taxonomies may struggle to capture relevant functions automation can be applied to. One-dimensional taxonomies can be criticized for not differentiating between the human information processing stages. Those taxonomies have problems when classifying, for instance, a technology with a high LoA for decision selection but a low level of support for action implementation, resulting in a lack of a corresponding unambiguous "level of automation".
2. Assigning precise LoA level. The differentiation between certain LoA may necessitate extensive knowledge in a specific domain, such as psycho-physiological expertise for assessing the levels of automation in a "perception task", where distinctions between the concepts of "sensation" and "perception" may be intricate. Additionally, for some taxonomies, these distinctions may not always be clearly defined (e.g., what is exactly the difference between "low automation" vs. "high automation"?), leading to difficulties in assigning and/or discerning between different levels or types of automation even among experts.
3. Defining "emerging task" when changing different LoA. Change in LoA may generate unforeseen behaviours and new task allocations not present in previous LoA. For instance, assigning a high LoA to task activity could introduce a new task, "supervise automation", which was not present at lower LoA, so this task was not previously assigned to the operator. This challenge becomes more pronounced in the presence of novel or undefined tasks, particularly when opaque and non-transparent artificial intelligence (AI) learning capabilities are involved in safety-related tasks to be supervised.
4. Characterizing complex and/or novel human-machine cooperation. Current taxonomies may fall short in adequately characterizing human-machine cooperation, often adopting a binary (man OR machine) view of automation (Dekker & Woods, 2002) rather than focusing on the question of "How do we make them effectively collaborate?" Clear responsibility and authority allocation are crucial to define roles and accountability while aligning the team's

perception, situation understanding, and future projections in a shared, transparent manner (Bouzekri et al., 2018).

5. Considering new concepts. The rise of AI-based technologies not only generates new emerging roles or cooperation modes but introduces new challenges for human-automation teaming (HAT) concepts. Especially with AI's capabilities that aim at playing the role of "virtual teammate", consolidated taxonomies and models are limited in describing how team members could establish common goals, communicate shared situation awareness, and become predictable to each other (Endsley, 2023).
6. Fully accounting for dynamically changing task allocation (adaptive and/or adaptable). The capability to dynamically allocate tasks between human and automation in complex and evolving environments necessitates a comprehensive and flexible framework able to allow modifications in LoA in response to real-time live events and human inputs (Johnson et al., 2014).

In light of these limitations, some researchers argue that there is no need to keep delving further into LoA concepts for the design of complex systems (Jamieson & Skraaning, 2018). Despite these critical points, LoA frameworks still continue to play a crucial role in systems engineering. This is because mapping of functional capabilities and allocating tasks between human and automation provides a fundamental conceptual basis for designing system interfaces. Engineers are still required to design and implement multiple concepts of interactions between humans and machines in collaboration with different stakeholders (e.g., legal and/or human factors experts) in order to produce effective interface design for functional distributions. Understanding what the human and computer/robot/virtual assistant are each doing and how they depend on each other, in a safe and effective manner, is still essential for effective system design (Ali et al., 2022).

5.1 GALA FRAMEWORK

Taking all these positive and negative factors into account, the human factors engineering team of AIRBUS Defence and Space GmbH, in Manching, has developed a framework called General Automation Level Allocation (GALA) framework (Table 5.1). The GALA framework employs a two-dimensional approach, striving for compatibility with previous frameworks while addressing some of their limitations. According to the framework, each task under analysis (generated by a hierarchical task analysis) will first be classified based on the four stages of the human information processing model presented in Parasuraman et al. (2000) as types of automation. For each of those stages, the taxonomy provides six LoA, ranging from 0: no support to 5: fully automated. This number of levels was chosen because it is still quite manageable for the analyst and at the same time offers sufficient granularity to distinguish them. By incorporating two dimensions (LoA and information processing stages), the structure is similar to other frameworks, e.g., the one by Save et al. (2012). However, the levels are defined in a generalizable way so that each level shares a common meaning across all stages. For example, an LoA of 0 always means that the human

TABLE 5.1
Levels of Automation according to the GALA Framework

Level/Stage	A. Information Acquisition	B. Information Analysis	C. Decision Selection	D. Action Implementation
0. No Support No or only low-tech artefact support	Operator acquires relevant information directly (not mediated via display), might include the usage of low-tech artefacts (e.g., binoculars).	Operator compares, combines, and analyses different information items, might include the usage of low-tech artefacts (e.g., pen & paper).	Operator generates decision alternative(s) and selects (chooses and confirms) the appropriate one.	Operator executes and controls all actions, might include the usage of low-tech artefacts.
1. Simple Support Simple function under control of operator	Sys acquires & provides information to operator.	Sys conducts "simple" analysis operations corresponding to lower level of cognitive skill (e.g., comparing information to detect differences, making linear predictions). Support is based on user's settings.	Sys generates & provides feasible decision alternative(s) and/or assists operator by generating decision alternatives (e.g., by evaluating them). Operator can select one of those, modify it, or generate and select an alternative option.	Operator initiates action implementation and is the prime executor. Sys executes part(s) of the action and/or provides execution support (e.g., by constraining action space). Operator is able to monitor, modify, and/or interrupt at any time.

(continued)

TABLE 5.1 (Continued)
Levels of Automation according to the GALA Framework

Level/Stage	A. Information Acquisition	B. Information Analysis	C. Decision Selection	D. Action Implementation
2. Complex Support Complex function under control of operator	Sys acquires & provides information to operator. Sys supports by integrating data, filtering and/or highlighting information. Support is based on user's settings.	Sys conducts "complex" analysis operations corresponding to higher level of cognitive skill (e.g., interpretation, classification, complex predictions). Sys might trigger alerts. Support is based on user's settings.	Sys generates & presents one or more feasible decision alternative(s). Sys preselects (chooses but doesn't confirm) one decision alternative. Operator can approve preselection or generate and select a different decision alternative.	Sys executes the full action sequence. Operator initiates action implementation or approves Sys initiation proposal. Operator is able to monitor, modify, and/or interrupt at any time. The operator must be ready to intervene (interrupt/adapt) immediately at any time.
3. Supervised Automation Automated function under supervision by operator	Sys acquires & provides information to operator. Sys supports by integrating data, filtering and/or highlighting information. Support is based on criteria set by sys or at design time. Criteria are still modifiable by the operator.	Sys conducts analysis operations corresponding to higher cognitive skill level (see above). Sys might trigger alerts. Support is based on criteria set by sys or at design time. Criteria are still modifiable by the operator.	Sys generates & presents one or more feasible decision alternative(s). Sys preselects one decision alternative. Operator has adequate time to veto before automatic decision implementation.	Operator or Sys initiates action implementation. Sys executes the full action sequence. The operator is able to monitor, modify and/or interrupt at any time. The operator must intervene within a reasonable time when prompted by the system.

4. Monitorable Automated Function Automated function sometimes checked by operator	Sys acquires & provides information to operator. Sys supports by integrating data, filtering and/ or highlighting information. Support is based on criteria which are set by sys or at design time. Criteria are visible/ accessible but not modifiable by the operator.	Sys conducts analysis operations corresponding to higher cognitive skill level. Sys might trigger alerts. Support is based on criteria set by sys or at design time. Criteria are visible/accessible but not modifiable by the operator.	Sys generates feasible decision alternative(s). Sys selects one of those autonomously (no further approval by operator needed). Operator is only informed about the decision.	Operator or Sys initiates action implementation. Sys executes the full action sequence. Operator can monitor and interrupt.
5. Fully Automated Function Automated function completely decoupled from any human intervention	Sys acquires information. Sys integrates and filters data. Support is based on criteria which are set by sys or at design time. Criteria are neither visible nor modifiable by the operator.	Sys conducts analysis operations corresponding to higher cognitive skill level. Sys might trigger alerts. Support is based on criteria set by sys or at design time. Criteria are not visible/ accessible and not modifiable by the operator.	Sys generates feasible decision alternative(s). Sys selects one of those autonomously (no further approval by operator needed). Operator is not informed about decision.	Sys initiates action implementation. Sys executes the full action sequence. Operator cannot monitor or interrupt action.

operator performs the function without any automation support. While the stages of the framework are drawn from the information processing model proposed by Parasuraman et al. (2000), GALA levels are primarily defined based upon thorough analysis of the following "one-dimensional" frameworks: the PACT framework, the "ten levels of automation for human–computer decision making" proposed by Sheridan et al. (1978) and the "levels of control and automation" by Endsley and Kaber (1999). The development also took into account the empirical experiences of the AIRBUS Human Factors Team, gained through the team's direct use of these frameworks in various projects over the years. In the context of system design, the GALA framework mandates the identification of relevant information processing stages (A, B, C, D) associated with the task or function. Then, GALA requires the determination of appropriate levels or a range of LoA for each implicated stage (0, 1, 2, 3, 4, 5). To facilitate this process, GALA provides a decision tree complemented by triggering questions (Ruscio et al., 2024).

5.2 EVALUATION OF THE GALA FRAMEWORK

Developing a new LoA framework in response to the shortcomings of existing ones is a common practice, as the multitude of existing frameworks testifies. Nevertheless, the novelty of a framework is no guarantee of its effectiveness in addressing known problems without a dedicated validation activity. Current literature on automation levels provides several methods that could be used to validate a framework. One way to evaluate an LoA framework is by applying it to various technologies to better understand its function, as it was done with the LOAT framework developed in the SESAR Joint Undertaking programmes. There, over 20 automation examples in Air Traffic Management (ATM) were analysed by Airspace Users and experts involved in SESAR projects, providing an empirical basis to assess the framework and derive design principles for ATM technologies (SESAR Joint Undertaking, 2013). Another approach involves laboratory studies, like those by Endsley and Kaber (1999), where participants performed dynamic control tasks to examine the effects of different LoA assignments on participants' performance, situation awareness, and workload. Systematic review of LoA is another approach, such as Vagia et al.'s (2016), which compared 12 LoA frameworks, highlighting strengths and weaknesses, and thus guiding the framework's evaluations. Bjurling et al. (2024) refined their HAT framework by applying it to different use cases and conducting semi-structured interviews with stakeholders. Similarly, Andersson (2008) studied the impact of different LoA concepts on human performance in nuclear power plants through observations and interviews, identifying LoA dimensions that affected operator-automation collaboration. In another study, Andersson (2011) evaluated LoA taxonomies across five domains—aviation, power, maritime, nuclear, and refineries—through field studies and operator interviews.

As briefly shown in this overview, there is no univocal way to "validate" a new LoA framework. For this evaluation study, it was decided to combine the SESAR's approach (applied technologies) with Bjurling et al.'s (experts' interviews) and Andersson's (different domains). The evaluation composed in this way focused on understanding how effectively GALA can be used for designing and classifying automated systems in different domains. It can be represented by two main questions:

1. "How can GALA be extended address open issues of automation in regard to human automation teaming, with focus on dynamic task allocation and human behaviour?"
2. "How can GALA support Human Factors engineers in different domains?"

To address the research questions, a series of interviews with LoA experts on LoA and practitioners from multiple domains were conducted. To address the first question, interviews were conducted to provide insights into open issues of automation and to understand how GALA might be used or adapted to meet the open issues highlighted by the experts. To address the second question, a second series of interviews with experts working in different domains were conducted to investigate how far GALA could be applied in use cases from different industries, such as automotive, robotics, and medicine.

5.2.1 Interviews: Open Issues and Application to Different Domains

Twelve semi-structured interviews were conducted in two series of interviews ("LoA Expert" and "Domain Expert" interviews) with two different main goals. The LoA Expert interviews focused on gathering comprehensive information about current LoA automation challenges (e.g., HAT, DTA etc.) and then on collecting the experts' comments about the GALA framework. The domain-expert interviews focused, instead, on: relevant use cases involving the design and assessment of automation; gathering information about which framework (if any) was the most used in their domain; and if the GALA framework could be possibly applied to design/assess the technologies they are currently developing in their domain of expertise. Similar to Bjurling et al. (2024), the qualitative interview format was chosen as a research method since it possesses the flexibility to adjust to the respective expertise of each participant, enabling deeper insights in certain topics, and further discuss them against the GALA framework.

5.2.1.1 Interview Sample

Six "experts in levels of automation" were recruited for the first series of interviews, based on their experience with LoA models or frameworks (inclusion criteria: having worked with, evaluated, developed, or even contributed to the creation of an LoA model/taxonomy/framework). The LoA experts that agreed to participate in the interview had backgrounds in aviation automation, with a median age of 48, three males and two females, based in France and Italy. Their current working titles were: Safety/Human Factors consultant for Aviation & Railway; Professor of Computer Science; Specialist for Automation & Robotics; Human Factors expert; and Head of Department Human Factors Engineering/UX Ui Design.

Six more "domain experts" were selected for their direct engineering experience with LoA in specific domains. The chosen domains for being included in the second series of interviews were: helicopters, robotics, medical technology, automotive industry, and architecture/urban planning. All experts were contacted via email. The domain experts were located in Germany, France, and the Netherlands, with a median age of 42.5, five males and one female. Their current working positions referred to: developer of driver assistance systems and automated driving systems; cockpit designer for helicopters; systems designer for architecture and urban planning applications; XR/VR ergonomist for chirurgic/medical instruments; trainer for technology company developing robots; and developers.

5.2.1.2 Interview Procedure

The first round of interviews focused on LoA experts. Before the interview began, participants were provided with an information sheet with the purpose of the study, and a consent form was obtained. After collecting demographic and professional background information, the initial phase of the interview involved general questions about automation. Potential challenges and concerns about the use of LoA models were discussed, with a particular focus on issues related to HAT and dynamic task allocation (DTA) concepts. In the second part, the interviewer introduced the GALA framework to the LoA expert, providing introductory materials and explaining the framework. Appropriate time to review the GALA framework was given, as well as the possibility to ask clarifying questions. Following this introduction, participants answered more specific questions about how the GALA framework could address HAT, DTA, and open issues that were mentioned in the first part of the interview. The requirements for a satisfactory framework, as defined by Howard (1997), were also discussed with the experts. Each part of the interview lasted around 40 minutes, making the total interview last approximately between one and two hours. The interviews were conducted online, and both recordings and transcripts were produced to support further analysis.

The second series of interviews focused on domain experts. Since the primary focus of these interviews was on applying the GALA framework to a specific example, domain experts were provided with introductory material about GALA framework in advance of the interview so that they could get ready to discuss it in relation to the automation concepts in their domain. Participants were asked to provide a use case or an example technology from their own domain to be analysed by GALA framework and to think about whether and how the GALA framework could be applied to that technology as part of their preparation before the interview. During the interview, the participants were provided with an information sheet and a consent form. After that, the demographic and professional background information was collected from the participants, followed with a brief recap of the GALA framework by the interviewer, after which any open questions could be asked by the participants. The core of the interview centred around applying the GALA framework to the selected example and discussing the challenges, limitations, and potential benefits that emerged from the use case(s) application. The interviews lasted between 30 minutes and one hour. As with the previous series, the interviews were conducted online. They were recorded and transcribed for subsequent analysis. All interviews were conducted in English or in German, and participants kindly agreed to contribute to this research on a voluntary basis and did not receive any monetary compensation for their participation in the interviews.

5.2.1.3 Interview Analysis

The interviews were analysed with thematic analysis (Braun & Clarke, 2012). Thematic analysis was chosen for the flexibility to be applied to a variety of qualitative data and the iterative nature of the approach. Furthermore, it is a robust method for small sample size. Content was coded, and patterns were identified. Emerged themes were reviewed and specified into a set of requirements. Data gathering and analysis was done on the software Miro (Figure 5.1). The final cluster of topics was then combined into main recommendations that were sent via email back to the expert for a final check on the overall results of the analysis, asking an opinion on what has emerged from the overall results.

5.2.1.4 Interview Results

Overall, all the experts agreed that the GALA framework can be used for designing and classifying automated systems. It presents some qualities that can address some open issues of automation level concepts, specifically on HAT and DTA. While many of the qualities recognized were already perceived as embedded in the framework, experts agreed that making these elements more explicit would enhance its clarity, for instance by making clearer the implications of each type/stage chosen. This could also support reasoning about how these aspects—such as roles and authority—evolve under different conditions. Greater flexibility through incremental extensions to an LoA framework was considered important, especially when dealing with complex or evolving situations, as it is almost impossible to account for every potential outcome or scenario in advance. However, the experts also raised concerns about how changes to include flexibility should be implemented within a framework like GALA. The experts noted, indeed, that these changes could make the framework more complex to use if changes are not carefully managed.

Some experts appreciated the complementary "decision tree" (visible in Ruscio et al., 2024) as a way to make the GALA framework more adaptable and responsive to dynamic scenarios, and thus suggested to include error models, such as Reason's General Error-Modelling System (GEMS), not only to address task allocation but also to include human reliability concepts during the system-design discussion. Other experts argued that this would unnecessarily complicate the framework and suggested that error analysis should be handled during the earlier task analysis phase and some of the triggering questions could not be immediately applicable at early stages of design.

Another key point raised by the experts was the need to clearly define the intended audience for the framework before implementing changes. It is crucial that users fully understand both the purpose of the framework and how to apply it correctly. One expert noted that the clarification of purpose and application alone may not be sufficient, stressing the need to clarify the underlying reasons for using LoA frameworks and how they should be applied, especially for users not familiar with human factors or task analysis methods.

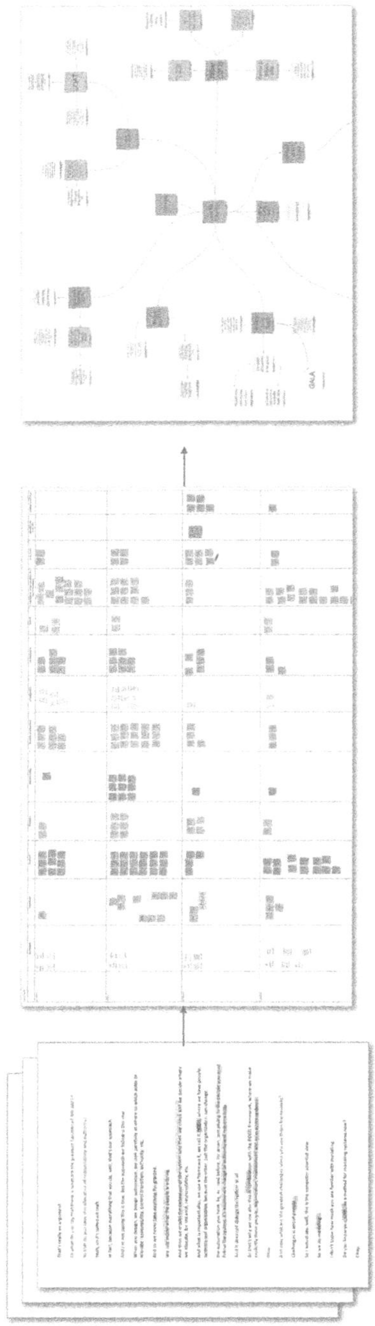

FIGURE 5.1 Process for thematic analysis used to analyse and cluster the results of the interviews.

Finally, some participants emphasized the importance of trying to prevent misinterpretations, by providing definitions or a glossary of words defining the two dimensions of the frameworks. A summary of the overall recommendations can be seen in Table 5.2.

As for the second research question (i.e., if the GALA framework could support human factors engineers in different domains), the GALA framework was considered to be applicable across all domains explored, though the extent of some caveats to be considered.

In the automotive industry domain, the SAE J3016 framework is the most widely recognized and established (SAE International, 2021). The challenges of trying to consider the GALA framework in automotive included difficulties distinguishing

TABLE 5.2
Overview on the Recommendations, Their Development, and Reflections on Their Implementation

Recommendations	1	2	3	4
Expert's Combined Input	Incorporate more flexibility into the framework by adding optional incremental extensions to better account for HAT. These have to be consistent and univocal.	Incorporate functions concerning responsibility, authority and control transition.	Extend the information processing stages to better represent human behaviour by incorporating other concepts and accounting for errors (by operator and/or automated system).	Facilitate the understanding of the framework by using clear terminology and providing descriptions and examples.
Open LoA Issues	HAT, DTA	Responsibility. Authority. Control transition.	Human behaviour performance. SA. Training. Errors.	Precision. Complexity.
Satisfactory Taxonomy Criteria (Howard, 1997)	Unambiguity. Exhaustiveness. Mutual exclusiveness.	Unambiguity.	Unambiguity. Exhaustiveness.	Unambiguity. Acceptability. Repeatability.
Implications for Implementation	Increasing complexity (positive and negative aspects). framework might not be the right place to address these issues.	Implicitly it is already included.	Increasing complexity (positive and negative aspects). Framework might not be the right place to address issues.	Might be insufficient.

between "information acquisition" and "information analysis". The automotive expert was, indeed, used to use the one-dimensional SAE framework where the driving task is divided into specific sub-tasks (e.g., steering, braking etc.), and clear responsibility allocation is referenced between the driver and the vehicle for the overall driving task. While the GALA framework offers a psychological perspective by incorporating information processing stages to the driving task (e.g., in order to "steer" information needs to be acquired and analysed before implementing the action of steering), automated driving solutions tend to involve more global shifts in automation, covering mainly the "action implementation aspects" rather than distinct steps. For this reason, the SAE J3016 framework was perceived as more practical for day-to-day engineering needs. However, a simplified version of GALA that focuses solely on action implementation was considered to be potentially useful in this context, particularly when considering automated assistance tools that impact decision-making aspects beyond the driving task itself, such as route choices or destination preferences.

In the medical technology domain, according to experts interviewed, there are no widely used LoA frameworks for medical technology. The GALA framework was discussed as applied to two examples: a spinal surgery robot and an automated system for tissue image analysis. In both cases, the framework was seen as valuable for guiding the design, development, and assessment of medical technologies. It offered a structured way to reason about how automation could evolve and what factors should be considered at each level. Since medical professionals remain cautious about embracing more automation, an incremental approach to automation was considered beneficial and recommended. The GALA framework could support this by helping manage the gradual transition, allowing personnel to adjust to and eventually accept the new automated systems. However, the framework does not currently explicitly take into account the varying degrees of acceptance or willingness of end users to adopt automation. Furthermore, the current technological and legal constraints in the medical field—especially in terms of safety and accountability—make the application of higher LoA unlikely with the current technology maturity level.

In the helicopter domain, participants were not aware of any specific LoA frameworks tailored for their technology. The work in this field tends to rely heavily on empirical research and direct interaction with end users. When the GALA framework was applied to a helicopter system that recognizes and reacts to obstacles, it was found to be effective by the interviewees. The decision tree, in particular, provided a useful classification for existing use cases and for the design phase. However, concerns were raised about the additional documentation required when using the GALA framework, with some participants viewing it as unnecessarily complex and redundant for their current needs.

In the robotics domain, the participant mentioned the "Levels of Collaboration (LoC) framework" as described by Bauer et al. (2016), but they had never used it in everyday practice. The GALA framework was successfully applied to two examples in the robotics domain, involving a robot that could navigate and conduct inspections autonomously, as well as a drilling robot. Despite its applicability, trust and safety concerns were identified as barriers to the implementation of the framework, especially when human safety is involved in complex and critical situations. The GALA framework was not supportive in improving the understanding about the way trust and safety concerns could be identified and discussed at different LoA.

Finally, in the field of architectural design no specific LoA frameworks were known to the participant. The design process in this sector is often broken down into three phases: analysis, synthesis, and evaluation. The first two steps, being highly creative, are seen as tasks that require human input. In this regard, a significant point that was raised focused on the need for users to understand how alternatives may be generated by automation and not "just" receive the information that new alternatives will be generated. In the evaluation phase, where decision support tools play a significant role, the GALA framework was seen as promising because the evaluation process holds some potential for automation. The GALA framework was successfully applied to a decision support tool, specifically in the decision-making stage. Information acquisition and action implementation were not deeply explored, though information analysis was recognized as potentially valuable, while information acquisition and action implementation were deemed less relevant for this domain. The decision tree, or an adapted version of it, could serve as a communication tool between designers and clients, helping to clarify whether a system is intended to simply visualize data or actively make decisions. Although the GALA framework was applicable to decision support systems, participants felt it is ultimately better suited for fully automated control systems.

Table 5.3 sums up the overall results in the different domains.

5.3 GALA FRAMEWORK FIRST EXTENSION

As the result of both LoA and domain-expert interviews, a series of recommendations and considerations were produced. An overview of the results can be found in Tables 5.2 and 5.3. To address the main point discussed with the experts, the GALA-O and guiding templates were developed. Their description will be provided in the following paragraphs. Additional expansion in reply to the full recommendations will be considered for the next development of the framework.

5.3.1 GALA-O for Dynamic Task Allocation

To address the challenges related to DTA, from the operator perspective, and take into consideration recommendations referring to framework complexity in certain domains (e.g., automotive context), a variation of the GALA framework has been developed and refined. The GALA-Operator (GALA-O) is a simplification of the GALA framework that is centred around the operator's role when adjusting the LoA settings at runtime to address and/or change automation support in complex and evolving environments. This means that—as operational conditions change or as the nature of tasks evolves—human operators can quickly adapt the LoA (of adaptive and/or adaptable systems) to best suit the current context by having the possibility to choose between simplified LoA. This capability ensures that the allocation of tasks remains responsive to shifting needs and remain compatible with the choices allowed

TABLE 5.3
Overview on the Application of the GALA Framework in Different Domains

Domains	Automotive	Robotics	Medical	Helicopters	Architecture
Use Case(s)	Takeover request of an SAE J3016 Level 3 automated vehicle to manual driving.	Inspection Robot, can navigate on its own. Drilling Robot.	Automated analysis of tissue images. Spinal Surgery Robot.	Automated landing with obstacle recognition & circumvention.	Automated planning support system.
Advantages of GALA Framework	Decision Tree facilitates the task. Appropriate for automated vehicles with lower LoA.	Provides an overview on what the other respective teammate of HAT can do. In general, useful for robotics.	Helpful for the design, development and judgement of medical technology. Has the potential to support incremental implementation of automation in the medical context. Can increase acceptance of the new technology in this domain.	Ensures consideration of all steps. Enables structured comparison of existing automated systems. Facilitates decision on how to proceed with current system. Decision tree rather supports analysis, some "guiding templates" models may be produced for design-level.	Can be used for communication with the client.
Disadvantages of GALA Framework	Complexity makes it unlikely to be used, especially in complex environments and DTA. Might be too detailed for the driver, since steps considered here are usually simpler.	Depending on context, the framework might not be used to its full potential. Lack of focus on trust (relevant in safety-critical domains).	Does not consider how ready the operator is to accept the new technology. Might not be used to full potential due to legal and safety constraints.	Means extra effort and documentation compared to current process.	The description of the information processing stages is not clear or not easily applied to the use case. It is not a tool for decision support.
Overall Applicability	Possible.	Possible.	Possible.	Possible.	Partially possible.

by system design, but it grants a more usable decision range to adjust during the critical situation to the desired level of support needed.

GALA-O simplifies the "information processing stages" dimension of the GALA taxonomy by condensing its stages into one synthetic dimension that combines "decision" and "action" stages together (Figure 5.2). The reason for not considering the first two steps is because it is assumed that during live/critical operations that trigger DTA, the operator may need to focus mainly on decisions and action implementations and may not need to reduce automation to perform manual information acquisition or information analysis. However, contrary to one-dimensional taxonomies, GALA-O still maintains the possibility to differentiate between the steps "decision selection" and "action implementation", if needed. It enables the operator to fine-tune live the support from automation to any given task, allowing them to modify independently the LoA in the decision-making and action implementation steps.

GALA-O also simplifies the "levels of automation" dimension by combining in one level the GALA levels 1–2 (that were distinguishing between "simple" and "complex" decision modifications and actions implementations). At the same time, GALA-O does not present the "level 0" from the full GALA table, because GALA-O is designed to address situations where automation is present. This means that situations that were designed to not include any type of support from the automation (i.e., level 0) cannot be dynamically re-allocated to automation in real-time decision if no automation support was envisaged. Similarly, "level 5" was not included because tasks designed at the "fully automated" level are typically not visible to the operator, or they are visible but designed so that there is no possibility to let the operator modify them during real-time operations.

The overall result of such simplification is that GALA-O is composed of three LoA in order to allow focusing on runtime changes in a more useful (i.e., by removing "levels of support" that are relevant only at a design stage) and usable way (i.e., compared to a complex table with 4 × 6 degrees of freedom possible). These three levels are: MONITOR (LoA 4), VETO (LoA 3), and EVALUATE (LoA 1 and 2). Each level in GALA-O provides distinct decision and action characteristics to meet the operator's evolving needs. If required, additional fine-tuning is still available, in case a split between the levels in decision and action support needs to be further specified by the operator. This means that every time a given task requires to take decisions or action, by using GALA-O operator can focus on the type of support desired, by just selecting one of the following three words.

EVALUATE: When it comes to decisions, an operator may evaluate decision alternatives generated by automated systems. Every time a given task requires to implement some actions, the operator can decide to evaluate how automation is implementing the action. In this collaborative level, the system generates decision alternatives or assists in action implementation and the operator can evaluate and modify them. The operator can approve decisions or initiate actions suggested by the system, maintain control during action execution, and intervene at any point to adjust decisions or stop actions.

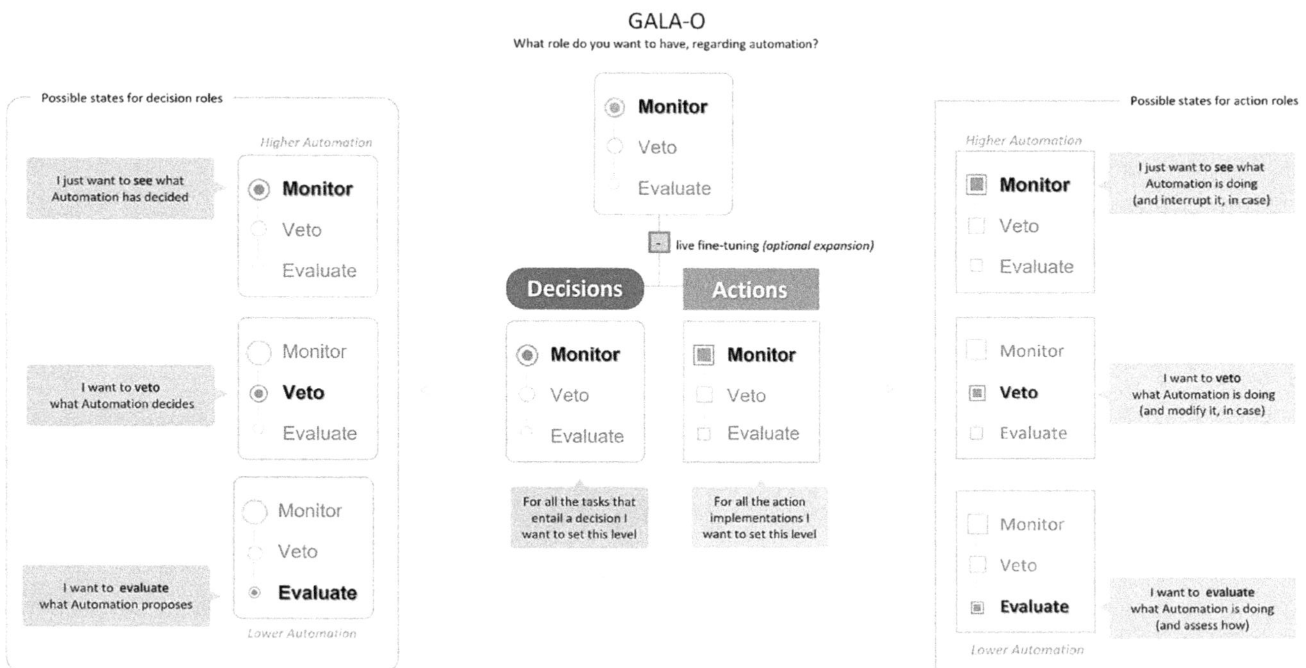

FIGURE 5.2 GALA-O widget and its state description for each level, for live operations.

VETO: The automated system decides, confirms, and executes the best option available, and the operator can, eventually, veto the options proposed or modify and interrupt the action being executed. In this level the system operates autonomously, but the operator can intervene and modify decisions and actions, if necessary. Automation handles decisions and actions under operator's control, which retains the authority to modify and veto decisions or actions and requests for modifications in case of incorrect or suboptimal outputs.

MONITOR: In this level automation decides and picks the best options, and operator only monitors what automation is deciding or is doing. In this level the operator mainly observes system decisions and actions without direct involvement, except for the possibility to interrupt the automated system. The system operates autonomously, handling decisions and executing and monitoring actions without requiring operator approval, but the operator is informed. This level is ideal when the operator's attention is needed elsewhere and the system is trusted and reliable. While the operator doesn't modify decisions or actions (like in the VETO level), they can interrupt the process, if needed, since it is possible to monitor automation.

It is important to notice that the operator or an assistance system may recommend to adopt any given GALA level, but not all levels are possible to be selected in case some boundaries were pre-determined (i.e., a range for LoA was set in accordance with legal, technical, and human factors considerations, done during the design phase). The GALA-O provides the operator with the possibility to dynamically change task allocation within the appropriate setting ranges made available (and compatible with the full GALA table), but GALA-O is not intended as a tool to override the LoA ranges defined at system-design time.

5.3.2 Guiding Templates for Discussing Roles and Emerging Tasks

From a system-design perspective, GALA requires to allocate LoA for each relevant activity derived from hierarchical task analysis or to automated functions that need to be designed or assessed. As emerged from the interviews, it is not always easy, however, to clearly anticipate how tasks, roles, performances, and responsibilities could change when selecting different LoA. Sometimes it is difficult to decide on a specific task allocation without anticipating how different LoA will impact workload and task complexity or will create new emerging tasks that could potentially impact performance and safety, thus potentially generating new requirements for automated systems at different LoA.

To support such activity, 24 "guiding modelling templates" based on UML models and SCXML standards were developed and refined. For all of the GALA's four "information processing" steps (A-B-C-D) and for all six "levels of automation" (0-1-2-3-4-5) within each step, a specific SysML activity diagram is created. The SysML activity diagrams used two swimlanes to create allocation relationships between a behaviour of action (e.g., the activities resulting from a hierarchical task analysis) and the blocks in which the action is contained (i.e., automation vs. operator). Each activity diagram represents the tasks allocated to "automation" (e.g., the engineered systems of interest) or to the human operator required to perform a given information processing stage in different levels of automation. Such representations allow us to visually translate the content of each "cell" of GALA table, showing not only the tasks allocated to different actors but also the behaviours that different LoA require to be considered for each information processing stage at different LoA. In addition, they allow us to model the information type to be displayed and transmitted (e.g., SysML "object") every time an action requires to "cross the swimlane" (i.e., from the operator to the system of interest, and vice versa).

For instance, Figure 5.3 shows the activity diagrams for two different cells of the GALA Table 1.1 for the support in the "C: Decision Making Process": level C1 vs. level C4. Even just by visually comparing them, it is possible to spot different philosophies in task allocation and potentially facilitate the discussion on related topics (e.g., workload, situation awareness, technical constraints etc.). In the first activity diagram (level C1), automation is assigned with two main activities: generate alternatives and display them to the operator. The operator is assigned with three main activities: perceive the displayed alternatives, assess which one is the optimal, and select and confirm the best option to conclude the decision process. In the second activity diagrams (level C4), it is possible to see how the same task can be modelled with higher LoA. At a level C4, the automation is in charge of not only calculating different alternatives and display them to the operator, but also assessing the best option possible and notifying the human operator about the decision taken. The operator can only acknowledge the result of the decision-making process. The comparison of the activity diagrams gives a first overview of: the task allocation and emerging task (e.g., "acknowledge" was not present at C1); the expected nominal taskload allocated to the human operator; and also, the different types of tasks involved (e.g., perception, cognition, responding task) at different LoA.

The scope of creating such templates to guide modelling activities is twofold:

1. Aid selection of automation levels associated with the respective information processing steps. The visual representation of the "template model" for each LoA can be useful to improve the understanding of the complex and multiple text-requirements that compose the "cells" of the GALA table. The UML models/SCXML standards should help the designer to model a given activity in different LoA, especially at early design stages, and enable the visual understanding of "emerging tasks" at different levels and stages. The abstract nature of the models should support the designer in focusing on the high-level differences among GALA levels, without lingering on technical details related to the feasibility of the contextualized activities that need to be modelled.
2. Support the understanding about the potential implications on human performance in interaction with automation. To account for the recommendations from the interviews with experts, such templates could support the visualization of tasks, objects, and information to be modelled from a system-design perspective. This may be particularly useful to anticipate, at early design

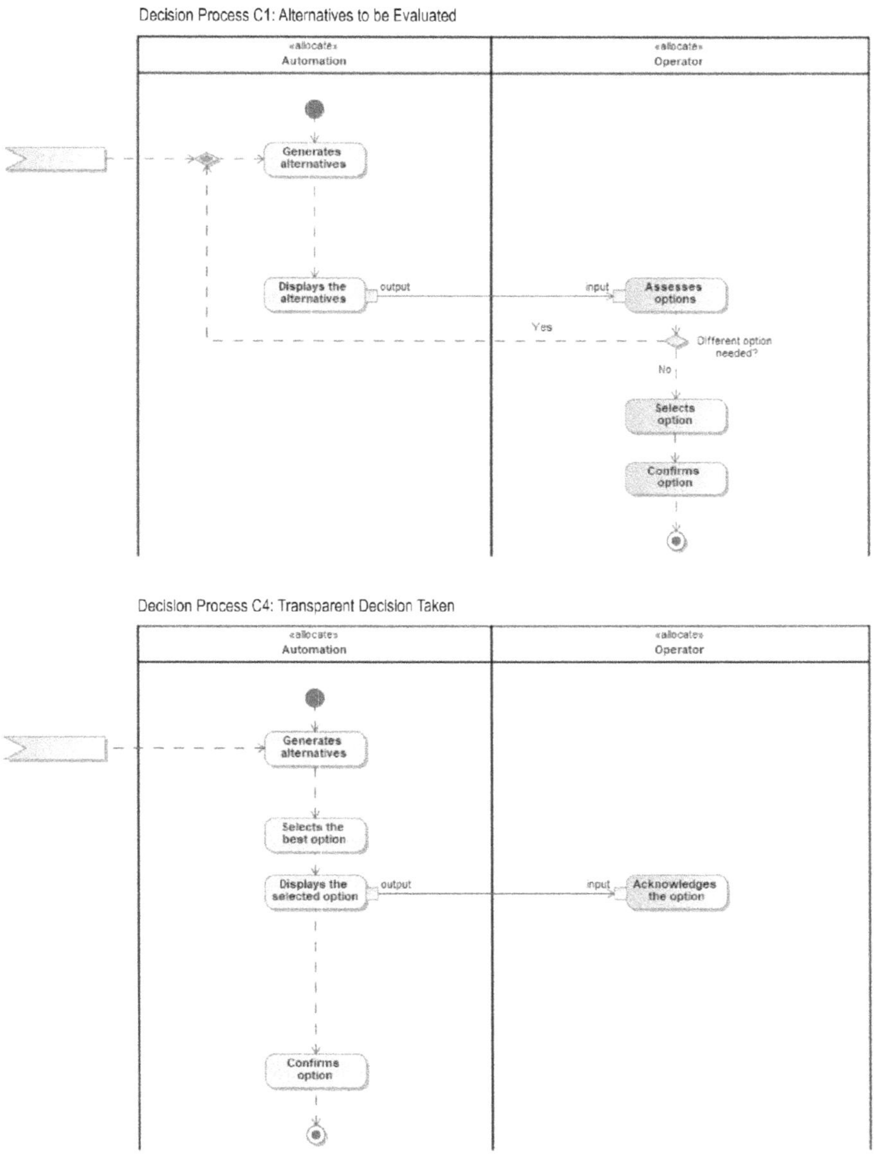

FIGURE 5.3 Examples of the template guide for task allocation modelling and assessing differences between two GALA levels for the decision process stage.

stages, the discussion about the implication of task allocation at different LoA, for instance with human factors engineers, UI developers, or legal experts. Such models could facilitate the discussions about: roles, responsibility, and legal considerations; generate requirements about which information (e.g., information types) the automation; and/or the operator may need to share and request in different moments of the activity diagram; and/or for discussing, for instance, situation awareness levels or taskload complexity at different stages and LoA.

Finally, it should be noted that the use of such SysML activity diagrams does not replace other existing methods or techniques for system design. Such “guiding templates” represent, instead, just a visual aid for the system designers and the engineers working on interface design, to better understand the difference in GALA levels and to support the considerations about how different LoA could change the nature of the tasks of the “human-automation team”. They can be used to interact with relevant stakeholders on the implications of modelling and/or describing the human-automation teamwork at different levels, but they do not aim to replace other assessment tools or design evaluation steps. Still, the guiding templates represent a useful standardized tool to decide the possible GALA levels and ranges allowed for each task at early design stages, when more complex taxonomies and frameworks cannot (yet) be fully applied (Bradshaw et al., 2013). Overall, the GALA guiding modelling template should reply to the recommendation about the need to support a better generation of a common ground and common understanding of automation among different stakeholders involved during the design or evaluation of task allocation and its related implications (e.g., emerging task, trust, workload, SA assessments, etc.) for HAT.

5.4 CONCLUSIONS

The evaluation activities conducted in this study highlighted that LoA frameworks are important and useful tools. The mapping and classification of tasks based on information processing stages and the consequent allocation of LoA are still required for the design of human-machine interaction concepts. Although this mapping may be complex, it does not necessarily mean that structured LoA frameworks cannot be used. On the contrary, it may be difficult to understand how to begin to identify the “things the human depends on the computer for” or “things the computer depends on the human for” if we do not precisely consider what things the human and computer (each) may be doing (Kaber, 2018). These insights collectively underline the necessity of considering a more dynamic, adaptable, and context-aware approach when applying LoA frameworks, recognizing the intricate interplay of psychological factors and the dynamic nature of real-world operational environments. As for the GALA framework, the overall results emerging from the evaluation with experts in different domains considered the new framework as potentially useful for matter and non-matter experts for assigning and discussing appropriate/required LoA and for enabling structured comparison of automated tasks and automation concepts. The GALA framework was considered able to fulfil most of the requirements for a framework as defined by Howard (1997) with potential for optimization concerning the use of the framework as well as its content. Participants, especially human factors experts, understood and accepted the GALA framework, indicating its potential to inform and structure automated system design. It was also considered theoretically applicable across all five domains studied.

However, some concerns and caveats arose during the interviews that can be summarized by the need to find a balance between flexibility (to be tailored to different domains, to be used by experts and non-experts) and complexity (to enable the discussion on automation-related topics). The GALA framework was considered to have some room for improvement in the definitions about when the framework should be used or when it should not, and why. This clarity could also help avoid critiques like those of Bradshaw et al. (2013) on autonomy assumptions. The recommendations collected contributed to a first extension of the GALA framework by offering a more realistic view of HAT and human behaviour, aiding the design of DTA. The introduction and finalization of GALA-O and the "guiding templates" could offer a structured approach to addressing some open issues in DTA and considerations on HAT respectively. The GALA-O facilitates adaptability in complex environments, by enabling adaptable automation level selection while reducing cognitive load and supporting more effective collaboration between humans and automation. The guiding templates could support the modelling activities and/or guide the discussion with different stakeholders, at the system-design level, about how human operators and automated systems work together efficiently with a shared understanding of roles, responsibilities, and emerging tasks.

It is believed that those first extensions of the framework could support the application of additional concepts and taxonomies (e.g., in the case of joint/shared human and automation activities) and could support the common ground between different stakeholders in discussing observability, directability, and predictability of automation. However, several limitations should be considered before consolidating these conclusions.

The main limitation of this evaluation is the small sample size, which affects the generalizability of the findings. However, these interviews used a qualitative approach, and providing quantitative indications about generalization of the GALA framework in specific domains was not the study's primary goal. The agreement between expert interview results supports the validity of the findings, though they can be further investigated by using different methodologies.

Future developments of the framework could focus on increasing flexibility by adding consistent incremental extensions to the GALA framework according to the whole set of recommendations. It can be important to clarify the transitions of responsibility, authority, and control and to enhance the representation of human behaviour in the information processing stages, incorporating errors from both operators and automated systems. Using clearer terminology, along with definitions and examples in a glossary, would facilitate better understanding of the framework. Lastly, integrating concepts such as embodied cognition, which links thought to physical actions and environmental interaction, could enrich the design of automated systems by accounting for not only cognitive aspects but also physical, sensory, and social factors involved in HAT considerations.

ACKNOWLEDGEMENTS

Thanks to Adrien Metge for his precious contributions. Research was supported by the Bundesamt für Ausrüstung, Informationstechnik und Nutzung der Bundeswehr under contract number E/L2FH/NA043/KF00.

REFERENCES

Ali, A., Azevedo-Sa, H., Tilbury, D. M., & Robert Jr, L. P. (2022). Heterogeneous human–robot task allocation based on artificial trust. *Scientific Reports*, 12(1), 15304.

Andersson, J. (2008). Levels of automation and user control-evaluation of a turbine automation interface (No. NKS–179). Nordisk Kernesikkerhedsforskning, Roskilde (Denmark)..

Andersson, J. (2011). Automation strategies in five domains: A comparison of levels of automation, function allocation and visualisation of automatic functions (No. NKS–237). Nordisk Kernesikkerhedsforskning, Roskilde (Denmark)..

Bauer, W., Bender, M., Braun, M., Rally, P. E. T. E. R., & Scholtz, O. (2016). *Lightweight robots in manual assembly—best to start simply* (p. 1). Frauenhofer-Institut für Arbeitswirtschaft und Organisation IAO.

Bjurling, O., Müller, H., Burgén, J., Bouvet, C. J., & Berberian, B. (2024). Enabling human-autonomy teaming in aviation: A framework to address human factors in digital assistants design. *Journal of Physics: Conference Series*, 2716(1), 12076.

Bonner, M., Taylor, R., Fletcher, K., & Miller, C. (2000). Adaptive automation and decision aiding in the military fast jet domain. In *Proceedings of human performance, situation awareness, and automation* (pp. 154–159).

Bouzekri, E., Canny, A., Martinie, C., Palanque, P., & Gris, C. (2018). *Using task descriptions with explicit representation of allocation of functions, authority and responsibility to design and assess automation.* In *IFIP WG 13.6 Working Conference Human Work Interaction Design (HWID 2018)*, 20 August 2018–21 August 20 (pp. 36–56).

Boy, G. A. (1998). Cognitive function analysis for human-centered automation of safety-critical systems. In *Proceedings of the SIGCHI conference on Human factors in computing systems—CHI '98*. ACM Press.

Bradshaw, J. M., Hoffman, R. R., Johnson, M., & Woods, D. D. (2013). The seven deadly myths of "autonomous systems". *IEEE Intelligent Systems*, 28(3), 54–61.

Braun, V., & Clarke, V. (2012). *Thematic analysis*. American Psychological Association.

Dekker, S. W. A., & Woods, D. D. (2002). Maba-MABA or Abracadabra? Progress on human-automation co-ordination. *Cognition, Technology & Work*, 4(4), 240–244.

Endsley, M. R. (2023). Supporting human-AI teams: Transparency, explainability, and situation awareness. *Computers in Human Behavior*, 140, 107574.

Endsley, M. R., & Kaber, D. B. (1999). Level of automation effects on performance, situation awareness and workload in a dynamic control task. *Ergonomics*, 42(3), 462–492.

Frohm, J., Lindström, V., Stahre, J., & Winroth, M. (2008). Levels of automation in manufacturing. *Ergonomia—an International Journal of Ergonomics and Human Factors*, 30(3).

Howard, J. D. (1997). *An analysis of security incidents on the internet 1989–1995*. Carnegie Mellon University.

Jamieson, G. A., & Skraaning Jr, G. (2018). Levels of automation in human factors models for automation design: Why we might consider throwing the baby out with the bathwater. *Journal of Cognitive Engineering and Decision Making*, 12(1), 42–49.

Janssen, C. P., Donker, S. F., Brumby, D. P., & Kun, A. L. (2019). History and future of human-automation interaction. *International Journal of Human-Computer Studies,* 131, 99–107.

Johnson, M., Bradshaw, J. M., Feltovich, P. J., Jonker, C. M., Van Riemsdijk, M. B. & Sierhuis, M. (2014). Coactive design: Designing support for interdependence in joint activity. *Journal of Human-Robot Interaction*, 3(1), 43–69.

Kaber, D. B. (2018). Issues in human–automation interaction modeling: Presumptive aspects of frameworks of types and levels of automation. *Journal of Cognitive Engineering and Decision Making*, 12(1), 7–24.

Lee, A., Baker, T. S., Bederson, J. B., & Rapoport, B. I. (2024). Levels of autonomy in FDA-cleared surgical robots: A systematic review. *NPJ Digital Medicine*, 7(1), 103.

Müller, R., Vette, M., & Geenen, A. (2017). Skill-based dynamic task allocation in human-robot cooperation with the example of welding application. *Procedia Manufacturing*, 11, 13–21.

Parasuraman, R., Sheridan, T. B., & Wickens, C. D. (2000). A model for types and levels of human interaction with automation. *IEEE Transactions on Systems, Man, and Cybernetics—Part A: Systems and Humans*, 30(3), 286–297.

Ruscio, D., Metge, A., Schopp, M., Dreyer, D., & Petermeier, B. (2024). The General Automation Level Allocation (GALA) Framework, or: Why do we need another level of automation framework? In I. L. Nunes (Eds.), *Human factors and systems interaction. AHFE (2024) international conference* (Vol. 154). AHFE Open Access.

SAE International. (2021). SAE Levels of Driving Automation™ Refined for Clarity and International Audience. www.sae.org/blog/sae-j3016-update

Save, L., Feuerberg, B., & Avia, E. (2012). Designing human-automation interaction: A new level of automation taxonomy. *Proceedings Human Factors of Systems and Technology*

SESAR Joint Undertaking. (2013). Guidance Material for HP Automation Support, 16.5.1, D04.

Sheridan, T. B., Verplank, W. L., & Brooks, T. L. (1978). Human/computer control of undersea teleoperators. In *The 14th annual conference on manual control.* NASA Ames Research Center.

Taylor, R. M. (2002). *Capability, cognition and autonomy. Proceedings of RTO human factors and medicine panel (HFM) symposium.* Advance online publication.

Vagia, M., Transeth, A. A., & Fjerdingen, S. A. (2016). A literature review on the levels of automation during the years. What are the different taxonomies that have been proposed? *Applied Ergonomics*, 53(Pt A), 190–202.

Yang, G.-Z., Cambias, J., Cleary, K., Daimler, E., Drake, J., Dupont, P. E., Hata, N., Kazanzides, P., Martel, S., Patel, R. V., Santos, V. J., & Taylor, R. H. (2017). Medical robotics: Regulatory, ethical, and legal considerations for increasing levels of autonomy. *Science Robotics*, 2(4), eaam8638.

6 Dumper and Excavator Operator's Workplace Risks and Ergonomic Issues

Vesna Spasojevic Brkić, Mirjana Misita, Ivan Mihajlović and Nemanja Janev

6.1 INTRODUCTION

The land, water, air, biota, and people are still impacted by the mining industry operations. Namely, the workplaces in mining are highly hazardous fields with a high rate of accidents, hazards, and disasters, which have a significant impact on the environment and human health [1–5]. In order to ensure that mining operations may run effectively, manage mineral exploitation responsibly, and prevent mining accidents or disasters, a number of acts, policies, and regulations pertaining to the mining industry have been developed [1, 6–7]. Despite the existing regulations, there are still serious risks associated with operating any heavy machinery [8–11]. Nonetheless, researchers studying heavy machinery often focus on the technical aspects of the system, neglecting the human operator as an essential part of the entire system [12, 13]. According to Refs. [12, 14, 15], incidents involving excavators and dumpers happen even more frequently than with other types of heavy machinery. Therefore, the purpose of this article is to survey possible workplace risks and ergonomic issues in dumper and excavator operations. The structure of this paper is as follows: after introduction previous research is presented, followed by methodology section, then the results are given, followed by conclusions containing future research avenues.

6.2 PREVIOUS RESEARCH

Operating an excavator and dumper is a physically and mentally demanding job [12, 15]. The excavator and dumper operators face a number of risks related to the machinery's operation and the working conditions. A well-designed, ergonomically adjusted cabin of a heavy vehicle not only improves the working conditions for the operator but also has a significant effect on site safety [16]. According to Eger et al. [17], just 6.4% of the mining industry and 15.8% of the construction industry self-report the implementation of ergonomic practices. According to Boudreau-Trudel et al. [18], improving ergonomics and safety in the context of heavy machinery is a very challenging task. According to Mandal and Srivastava [19], dumper operators

DOI: 10.1201/9781003672036-6

experience whole-body vibration and acceleration ranging from 0.644 to 1.82 m/s^2. As a result, they experience a lower quality of life, as well as lower back pain (85%), neck pain (37.5%), shoulder pain (30%), and ankle pain (37.83%). According to Upadhyay et al. [20], 58–74% of dumper operators had high and medium degrees of musculoskeletal disorders. Operators of excavators reported musculoskeletal problems in 76% of cases and blamed vibrations, emotional tension, or awkward posture at work for their problems [21]. Ramar et al. [22] discovered that dumper operators are more likely to experience neck and back pain as a result of the demand for a high number of operation cycles. An awkward postural requirement, such as static sitting, may arise for operators due to inadequate cabin design or unacceptable working procedures while they are carrying out their job responsibilities in the cabins [23–27]. Lin [28] observed that a larger range of adjustment for seat characteristics and an improved mechanism for seat adjustment can satisfy a wider number of operator requirements, which in turn reduces discomfort to body parts and increases subjective preference. Operators may be exposed to the following risks in addition to others: extreme temperatures, time constraints, dust, exhaust gases, psychosocial factors, whole-body vibration, and shift and prolonged work [25, 29–34]. Furthermore, the repetitious nature of the activities and the operator's posture while sitting are contributing factors to their discomfort [24, 35–36]. In addition, multiple research studies have shown that operators experience whole-body vibration, which, when combined with a hunched posture, exacerbates communication issues, lower back pain, and resonance excitation of the person [25, 37–40]. Operator posture is evidently influenced by cabin design, work needs, and anthropometric measures of the operators [16, 26, 40]. Backrests and armrests are unlikely to be intentionally used even when present while in an unpleasant position [16, 40, 41].

Nevertheless, it is uncommon to find checklists for evaluating the cabin design in the literature. Spasojevic Brkic et al. [9, 40] recently offered checklists for evaluating the cabin design. It is evident that research in the field of this paper is rare, so this survey aims to fill the evidence gap.

6.3 METHODOLOGY

Methodology included data collection on mining sites in Serbia and Montenegro from May 2023 to October 2024 by the questionnaire/checklist, by taking anthropometric measurements of operators, and by using measurement equipment. Further analysis on data collected by the questionnaire/checklist included descriptive statistics and hypothesis testing. Further analysis on data collected by taking anthropometric measurements of operators included 5th and 95th percentile calculations. Further analysis on data collected by using measurement instruments included collected data comparison with available regulation documents – directives and standards. The list of measurement instruments is shown in Table 6.1.

The three measurement points on mining machinery, as shown in Figure 6.1, were determined. Vibration of the whole body is measured with a sensor attached to the operator's seat. Indoor work conditions (air quality, gases, ppm, and noise) are measured above the operator's head in such a way that it does not obstruct visibility

TABLE 6.1
List of Used Measurement Instruments

Measurement Instrument	Measured Parameter	Calculated Parameter
TROTEC PC 220	Particle size: 0.3 μm 0.5 μm 1.0 μm 2.5 μm 5.0 μm 10 μm AT (°C)	• Permissible value in relation to the 24-hour PM2.5 standard • Permissible value in relation to the annual PM2.5 standard
HUMAN-VIBRATION ANALYZER VM31	Move on three orthogonal axes: aw_x aw_y aw_z aw(Vec) MTVV	• Daily exposure limit - eight hours; A(8) • Daily action value of exposure - eight hours; A(8)
DIGITAL ALMEMO D6 SENSOR/AHLBORN Universal measurement instruments and data loggers ALMEMO 2590-2A/4AS	CO_2 [ppm] LV NO_2 [ppm] LV SO_2 [ppm] CO_2 [mg/m^3] LV NO_2 [mg/m^3] LV SO_2 [mg/m^3]	• Threshold values • Short-term limit values (LVs)
HD2010UC/A Spectrum Analyzer Integrating Sound Level Meter	LC_{pk} – peak sound pressure value LA_{eq} – equivalent noise level LA_{Fp} – sound pressure	• p_{eak} value of sound pressure (maximum value "C" of frequency-weighted current sound pressure) • $L_{A,EX,8h}$ level of daily exposure to noise (dB(A) in relation to the reference value of 20 μPa) • $L_{A,EX,15min}$ level of 15 minutes of noise exposure • $L_{A,EX,8h}$ level of weekly exposure to noise
TESTO 340 FLUE GAS ANALYSER FOR EMISSION MEASUREMENT	ppm CO ppm NO % CO_2 °C FT ppm NO_x	• Allowed values in relation to STAGE I–V, year of engine production and engine power

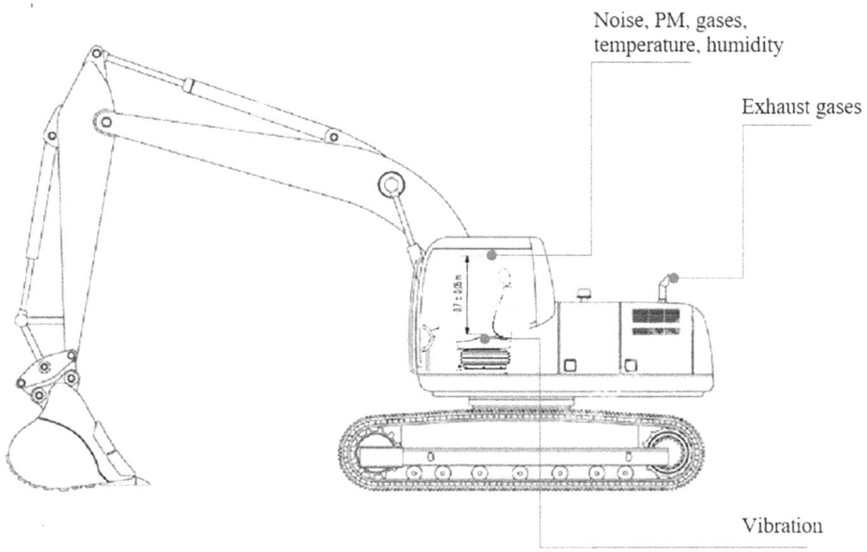

FIGURE 6.1 Measurement points at excavator.

or disturb the operator in any other way. The measurement instrument for the measurement of exhaust gases is placed on the third measurement point.

The regulation documents [42–52] were used to compare collected data. Finally, the list where measured and prescribed values of measured parameters have to be recorded and compared was designed, as in Figure 6.2, where the recording sheet is presented.

6.4 RESULTS

During the research period a total of 120 excavator and dumper operators were surveyed. There were 73 excavators and 47 dumper operators in the sample. All participants were male. First, the basic data of the operator was recorded, namely: age, height in cm, weight in kg, and years of work experience. Table 6.2 shows the descriptive statistics for excavator and dumper operators separately, and the data do not indicate significant differences in all observed values.

Data are collected also about anthropometric measurements of operators, and 5th and 95th percentile values are calculated, as in Table 6.3. Anthropometric measurements were collected and processed on the sample of 50 operators who agreed to participate in the measuring process.

For the purposes of the research, a checklist was used. The original checklist from research [9, 40] contained 39 of the 62 questions shown in Table 6.4, in such a way that the survey was expanded on the basis of the previous research discussed earlier in this chapter. Also, for 50 surveyed operators (whose anthropometric measurements are shown in Table 6.3), Table 6.4 shows mean values and standard deviations classified

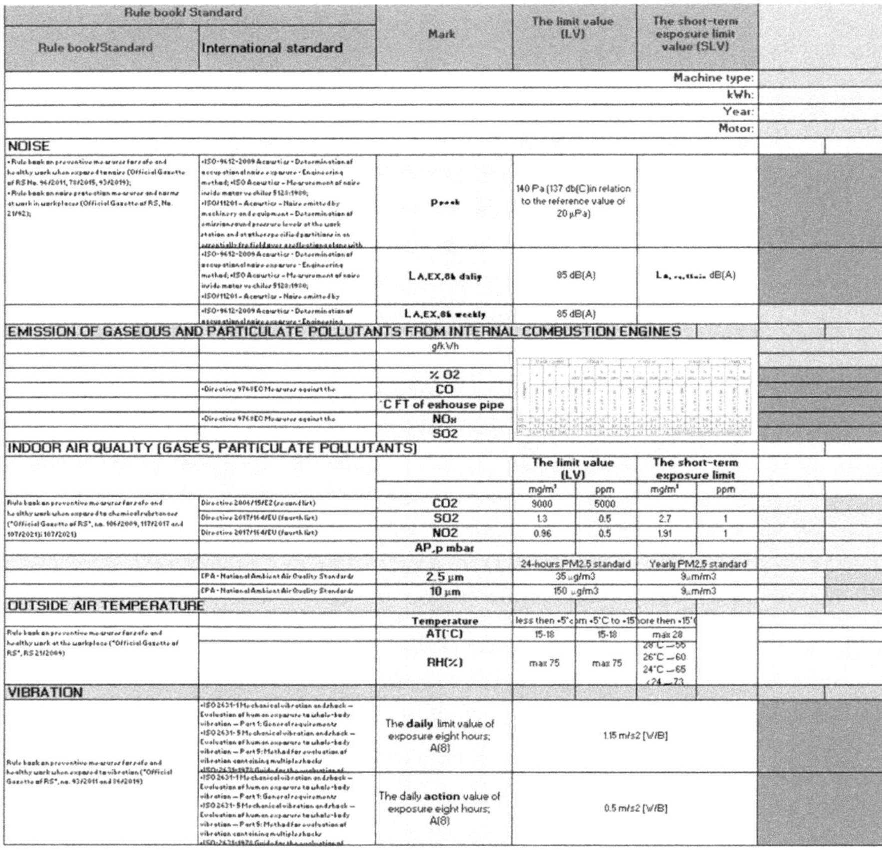

FIGURE 6.2 Recording sheet.

TABLE 6.2
Descriptive Statistics for Dumper and Excavator Operators Separately

	Excavator Operators				Dumper Operators			
Statistic	Age	Height (cm)	Wight (kg)	Years of Experience	Age	Height (cm)	Wight (kg)	Years of Experience
No. of observations	73	73	73	73	47	47	47	47
Minimum	23.0	162.0	70.0	1.0	19.0	166.0	60.0	1.0
Maximum	62.0	195.0	115.0	37.0	54.0	202.0	150.0	38.0
Range	39.0	33.0	45.0	36.0	35.0	36.0	90.0	37.0
Median	40.0	181.0	87.0	13.0	36.0	180.0	90.0	10.0
Mean	40.8	180.6	88.2	14.7	35.8	180.1	92.9	11.7
Variance	88.7	42.4	107.9	107.0	78.0	50.9	353.5	90.6
Standard deviation	9.4	6.5	10.4	10.3	8.8	7.1	18.8	9.5

by excavator and dumper operators and the Mann–Whitney test (Asymp. Sig. 2-tailed test) for comparison of data.

Table 6.4 shows that significant differences have been found regarding seat ergonomic characteristics including armrests, location and usage of controls/levers and pedals, size of the cabin and entrance possibility, discomfort/pain in head and trunk, visibility and general view inside and outside the cabin, control of temperature and

TABLE 6.3
Anthropometric Measurements of Operators (5th and 95th percentiles)

Anthropometric Measurement	5th Percentile	95th Percentile
Height (cm)	170	192
Eye height (cm)	151.6	180
Shoulder height (cm)	142.8	161.2
Lower arm height (cm)	71.6	81.2
Horizontal arm reach (cm)	66	80
Hip width (cm)	32	42.55
Sitting height (cm)	83	135
Eye sitting height (cm)	73	124.1
Lower leg height (cm)	53	60.65
Upper leg length (cm)	45.9	65.55
Shoulder width (cm)	39	49.55

TABLE 6.4
Mean Values, Standard Deviation and Mann–Whitney Test of Responses Received from Excavator and Dumper Operators'

	Excavator Operators		Dumper Operators		Mann–Whitney Test
	Mean	Standard deviation	Mean	Standard deviation	Asymp. Sig. (2-tailed)
1. Is the seat height adjustable?	3.73	1.311	5	0	**0.012**
2. Can the seat be adjusted horizontally?	3.43	1.331	5	0	**0.003**
3. Is the seat set at the appropriate height?	3.77	1.165	5	0	**0.009**
4. Does the seat have back support?	3.7	1.264	5	0	**0.009**
5. Does the seat have lumbar support?	3.53	1.383	4.6	0.894	0.074
6. Are there armrests?	3.87	1.106	2.2	1.789	**0.040**
7. Are the armrests adjustable?	3.77	1.135	2.2	1.789	**0.049**

(continued)

TABLE 6.4 (Continued)
Mean Values, Standard Deviation and Mann–Whitney Test of Responses Received from Excavator and Dumper Operators'

	Excavator Operators		Dumper Operators		Mann–Whitney Test
	Mean	Standard deviation	Mean	Standard deviation	Asymp. Sig. (2-tailed)
8. Are the armrests set at the appropriate height?	3.83	1.02	2.4	1.949	0.097
9. Is the seat firmly attached to the cab floor?	4	1.114	5	0	**0.030**
10. Can the seat be tilted back?	3.9	1.094	4.2	1.789	0.293
11. Can the seat rotate?	2.53	1.432	1.8	1.789	0.226
12. The seat can be folded sideways.	3	1.414	3.8	1.789	0.227
13. The vibrations that are part of the working conditions bother me.	3.23	1.524	3.4	1.673	0.808
14. Do you feel vibrations over the seat?	3.23	1.331	3	1.414	0.642
15. Do you feel vibrations over the floor?	3.3	1.418	3.4	1.673	0.865
16. Do you feel the vibrations through the controls?	2.8	1.4	2.4	1.342	0.546
17. I feel strong vibrations over my hand.	2.9	1.447	2.4	1.342	0.423
18. I feel strong vibrations over my whole body.	3	1.438	2.8	1.095	0.681
19. The position of the controls/levers and displays is appropriate.	3.93	0.98	3.8	1.789	0.767
20. Can the location of the controls or levers be adjusted?	3.13	1.383	4.8	0.447	**0.007**
21. Can you easily reach the controls or levers?	4.1	0.885	4.8	0.447	0.085
22. Can you easily move the controls or levers?	4.07	1.048	5	0	**0.029**

TABLE 6.4 (Continued)
Mean Values, Standard Deviation and Mann–Whitney Test of Responses Received from Excavator and Dumper Operators'

	Excavator Operators		Dumper Operators		Mann–Whitney Test
	Mean	Standard deviation	Mean	Standard deviation	Asymp. Sig. (2-tailed)
23. Can you easily reach the pedal?	4.27	0.868	5	0	**0.047**
24. Can you use the pedal easily?	3.97	1.159	5	0	**0.038**
25. Is the cabin large/ spacious enough for you?	3.67	1.184	5	0	**0.013**
26. The cab can be translated or rotated to ensure optimal visibility.	2.83	1.392	2.6	2.191	0.826
27. The equipment in the cabin has thresholds.	2.9	1.398	4.2	1.095	0.065
28. The equipment has handrails.	3.67	1.213	4.6	0.894	0.089
29. Can you easily open/ close the cabin door?	3.87	1.074	5	0	**0.012**
30. Can you easily get in/ out of the cab?	3.7	1.236	4.8	0.447	**0.043**
31. Do you have the proper equipment to enter the cabin?	3.7	1.317	5	0	**0.013**
32. Do you have the proper equipment to get out of the cabin?	3.67	1.295	5	0	**0.009**
33. I feel discomfort in my body as a result of reduced comfort in the cabin.	2.23	1.524	3	2	0.372
34. I feel discomfort in certain part(s) of my body as a result of reduced comfort in the cabin.	2.17	1.44	3.6	1.673	0.073
I feel discomfort/ pain in the area as a result of reduced comfort: [hand]	1.63	1.066	3	2	0.102
I feel discomfort/pain in the area as a result of reduced comfort: [leg]	1.7	1.208	3	2	0.100

(continued)

TABLE 6.4 (Continued)
Mean Values, Standard Deviation and Mann–Whitney Test of Responses Received from Excavator and Dumper Operators'

	Excavator Operators		Dumper Operators		Mann–Whitney Test
	Mean	Standard deviation	Mean	Standard deviation	Asymp. Sig. (2-tailed)
I feel discomfort/pain in the area as a result of reduced comfort: [had]	1.43	0.935	3.4	1.673	**0.004**
I feel discomfort/ pain in the area as a result of reduced comfort: [trunk]	1.57	1.04	3.4	1.673	**0.010**
I feel discomfort/ pain in the area as a result of reduced comfort: [neck]	2	1.39	3.4	1.673	0.063
I feel discomfort/ pain in the area as a result of reduced comfort: [back]	2.1	1.626	3.4	1.673	0.088
35. Do you have good visibility and general view of the work area?	3.97	0.89	5	0	**0.012**
36. Are the cabin windows without distraction?	3.63	1.098	4.4	1.342	0.100
37. Is there a device that allows better visibility of the working field?	3.6	1.221	3.6	1.673	0.845
38. Adequate visual or auditory signals are provided that inform me of the occurrence of an unsafe situation and the need to urgently stop the operation that has started.	3.23	1.305	4	1.732	0.168
39. The lighting in the cabin in night conditions and with reduced visibility is sufficient for unobstructed monitoring of information from all instruments in the cabin.	3.7	1.179	4.8	0.447	**0.032**

TABLE 6.4 (Continued)
Mean Values, Standard Deviation and Mann–Whitney Test of Responses Received from Excavator and Dumper Operators'

	Excavator Operators		Dumper Operators		Mann–Whitney Test
	Mean	Standard deviation	Mean	Standard deviation	Asymp. Sig. (2-tailed)
40. During work, I am in a position to notice warning signals from devices in the cabin.	3.73	1.112	4.8	0.447	**0.034**
41. I am bothered by the temperature and humidity that are part of the working conditions.	2.93	1.413	3.6	1.673	0.308
42. I can control the temperature and humidity in the cabin.	3.23	1.251	4.4	0.894	**0.049**
43. Do exhaust gases and dust bother you?	2.7	1.442	3.8	1.789	0.126
44. Do you mind pollution that is part of working conditions?	2.83	1.44	3	1.871	0.716
45. The presence of dust in the air in the cabin bothers me.	3.03	1.402	3	1.871	0.942
46. Sometimes the level of gases or dust in the air in the cabin is such that it affects the difficult handling of the load or even requires a temporary stoppage of the operation.	2.87	1.332	2.6	2.191	0.717
47. Occasionally, while I'm at work, I feel a cough, headache, and shortness of breath.	2.9	1.423	3	1.581	0.885
48. I have a feeling that in the cabin is noisy.	2.87	1.306	3	2	0.866
49. I feel the intensity of noise in the cabin such that it affects the identification of sound and warning signals or communication with other people.	2.73	1.258	3.4	1.817	0.322

(continued)

TABLE 6.4 (Continued)
Mean Values, Standard Deviation and Mann–Whitney Test of Responses Received from Excavator and Dumper Operators'

	Excavator Operators		Dumper Operators		Mann–Whitney Test
	Mean	Standard deviation	Mean	Standard deviation	Asymp. Sig. (2-tailed)
50. I notice a change in the volume and sharpness of the noise in the cabin (not only its intensity).	3.13	1.358	3	2	0.942
51. I feel a strong volume and sharpness of the noise in the cabin (not only its intensity).	2.97	1.245	3.2	2.049	0.575
52. The noise that is part of the working conditions bothers me.	2.83	1.262	2.8	1.789	1.000
53. Hygiene inside the cabin is regularly maintained (cleaning, degreasing, etc.).	3.8	1.064	4.4	0.894	0.237
54. The cabin filter is cleaned regularly.	3.77	1.223	4.6	0.894	0.123
55. Cabin windows are regularly kept clean of dust and other conditions of the external working environment (rain, ice, snow...).	3.77	1.165	4.6	0.894	0.093
56. Fire and explosion prevention measures are regularly implemented in the machine cabin.	3.87	1.074	4.8	0.447	0.058
57. I use personal protective equipment as an additional protection measure during work.	3.6	1.192	4	1.732	0.294
58. I feel a lot of stress at work.	3.17	1.262	3.2	1.789	0.922
59. Due to poor working conditions, I am often absent from work (sick leaves).	2.97	1.497	3.2	2.049	0.626

TABLE 6.4 (Continued)
Mean Values, Standard Deviation and Mann–Whitney Test of Responses Received from Excavator and Dumper Operators'

	Excavator Operators		Dumper Operators		Mann–Whitney Test
	Mean	Standard deviation	Mean	Standard deviation	Asymp. Sig. (2-tailed)
60. The atmosphere of cooperation and togetherness prevails among the operators.	3.57	1.251	5	0	**0.010**
61. Managers motivate and reward us.	3.4	1.192	3.6	1.342	0.826
62. Machine failures are very often caused by human and organizational factors.	3.2	1.215	3	2	0.903

TABLE 6.5
Description of Observed Mining Machines

Type	Open Pit Mine Type	Date of Recording	Engine Power (KW)	Year of Manufacture
Excavator #1	Surface mining of quartz sand	30.5.2024	123	2018
Excavator #2	Surface mining of quartz sand	27.6.2024	180	2024
Excavator #3	Surface mining of zinc and lead	29.7.2024	370	2017
Excavator #4	Surface mining of zinc and lead	30.7.2024	370	2017
Excavator #5	Surface mining of marble and granite	20.9.2024	220	2023
Dumper #1	Surface mining of zinc and lead	29.7.2024	250	2008
Dumper #2	Surface mining of marble and granite	20.9.2024	298	2010

humidity, and the atmosphere of cooperation and togetherness that prevails among the operators.

Important factors of the workplace conditions, of the operators on excavator and dumper mining machinery, were measures out during 2024 at seven workplaces. Of which five workplaces were on excavator while two were on dumper units. Table 6.5

presents the type of open pit mining, recording date of the working machine observation, engine power, and its year of manufacture.

Figure 6.3 shows a continuous view of noise measurements at the workplaces of operators on the observed excavator and dumper mining machines. The horizontal lines in Figure 6.3 represent the permissible limit values for the peak value of sound pressure (P_{peak}), which is 140 Pa (137 db(C) in relation to the reference value of 20 μPa), and the level of daily exposure to noise $L_{A,EX,8h}$ (dB(A) in relation to the reference value of 20 μPa), which is the time-weighted mean value of the level of exposure to noise for an eight-hour working time, and the limit value is 85 dB(A).

As can be seen from Figure 6.3, during the monitoring of the measured values for the sound pressure level and the equivalent eight-hour noise level, no exceedances of the permitted limit values were recorded at all observed workplaces.

Figure 6.4 shows the result of measuring temperature – AT (°C) and humidity – RH (%) inside the operating cabin. As the limit values for AT (°C) and RH (%) are calculated in relation to the air temperature in the external environment, Figure 6.4 shows the measurement results of all three quantities. When the air temperature in the external environment is higher than +15°C, the permissible temperature limit in the cabin space is 28 (°C), and the humidity is 55%. In Figure 6.4, it can be seen that with all recorded data from the field, the temperatures of the outside air were above +15°C, and therefore a limit value (horizontal line) of 28 (°C) was entered for the temperature in the cabin space. It can be noticed that it was exceeded in the Excavator #1, #2, #3, and Dumper #1, #2. The limit value for air humidity was not exceeded at all recorded workplaces.

Figure 6.5 shows the results of measuring the vibrations of the entire body in the observed mining machines. The vibrations of the entire body were measured at the operator's seat using the measurement instrument VM31, which, based on the displacement along the x, y, and z axes, calculates the vector A(8), which has a permissible limit value of 1.15 m/s^2 [W/B] for an eight-hour daily exposure to vibrations. This limit value is not plotted in Figure 6.5 because the measured values of the velocity A(8) for the eight-hour exposure of the whole body to vibrations for all observed mining machines are below 0.53 m/s^2 [W/B]. The same measured values from the measurement sensor for whole-body vibrations are also used to evaluate the action values of the eight-hour exposure to vibrations, whose limit value of 0.5 m/s^2 [W/B] is plotted in Figure 6.5. The results of the measurement of the action values of the eight-hour exposure to vibrations were exceeded at two workplaces (Excavator #5 and Dumper #1), as shown in Figure 6.5.

Figure 6.6 shows the results of measuring the particulate matter ($PM_{2.5}$) concentration in the air. According to the EPA – National Ambient Air Quality Standards (40 CFR part 50), the 24-hour limit value for $PM_{2.5}$ is 35 μm/m^3, which is indicated in Figure 6.6 with the horizontal line. The results of the measurements on the observed excavators and dumpers indicate that only Excavator #1 exceeded the permitted limit value.

Figure 6.7 shows the results of measuring particulate matter (PM_{10}). According to EPA – National Ambient Air Quality Standards (40 CFR part 50), the 24-hour limit value for PM_{10} is 150 μm/m^3, which is indicated in Figure 6.7 with a horizontal line. The results of the measurements on the observed excavators and dumpers indicate

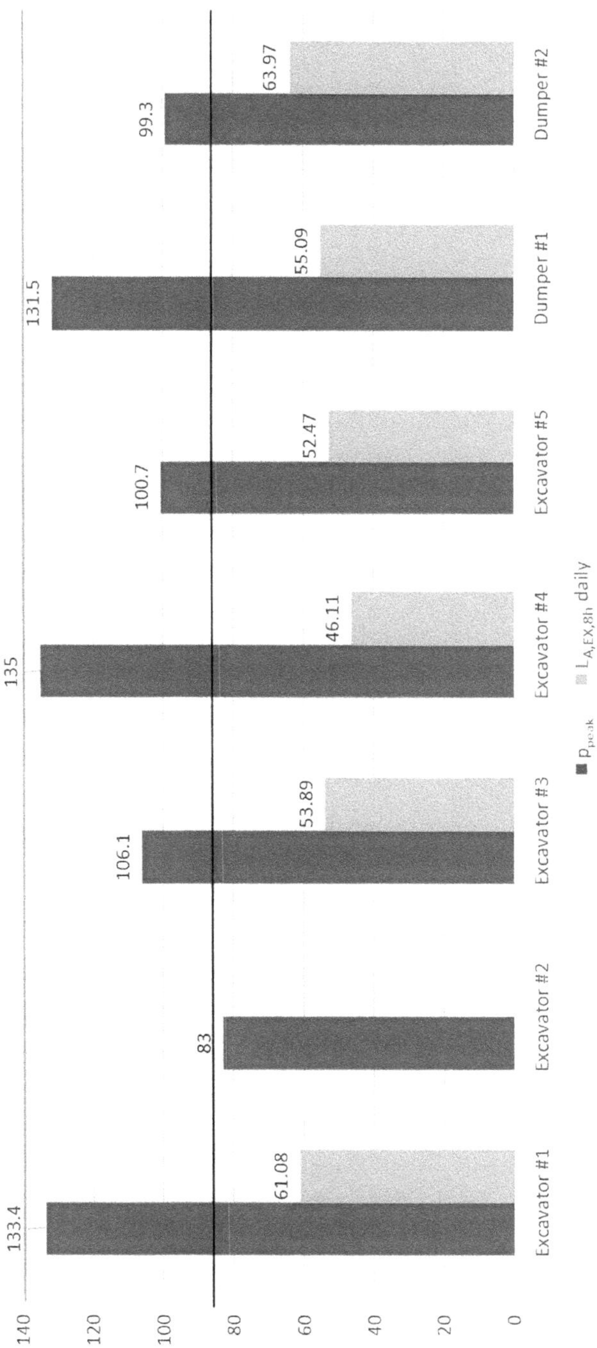

FIGURE 6.3 Recorded values of the peak value of sound pressure – P_{peak} and the level of daily exposure to noise – $L_{a,ex,8h}$, daily at the observed workplaces.

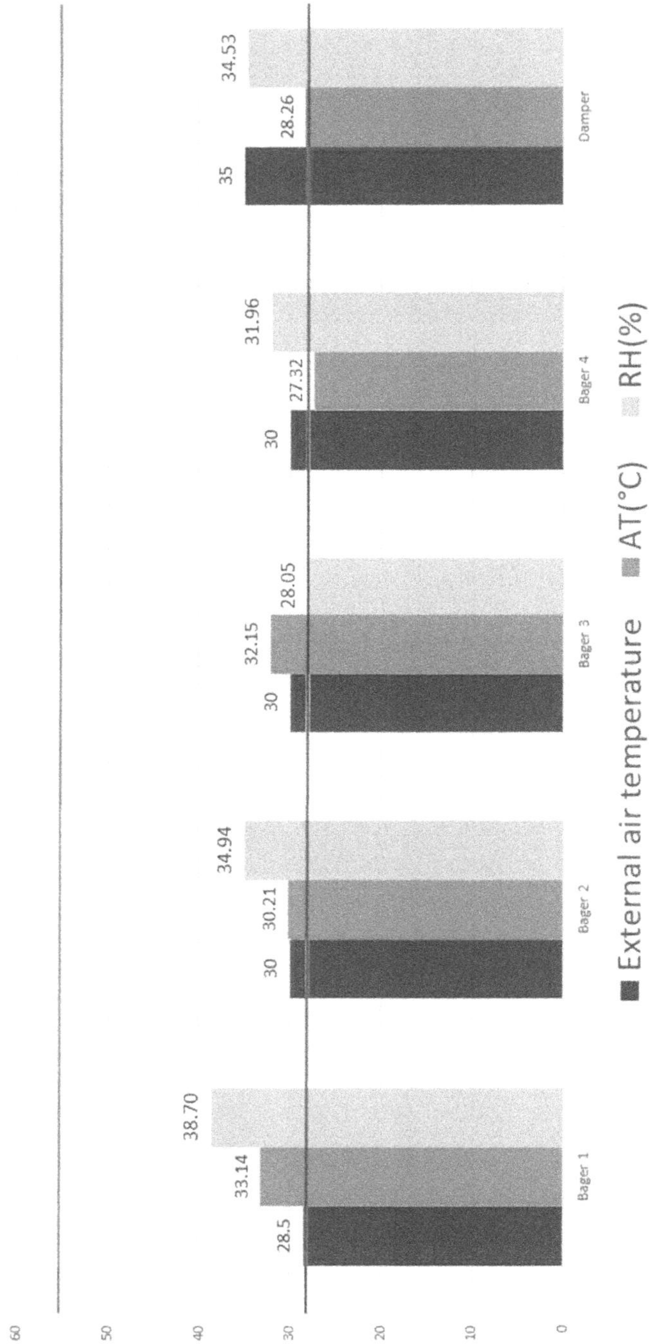

FIGURE 6.4 Recorded values of temperature – AT (°C) and humidity – RH (%) at the observed workplaces.

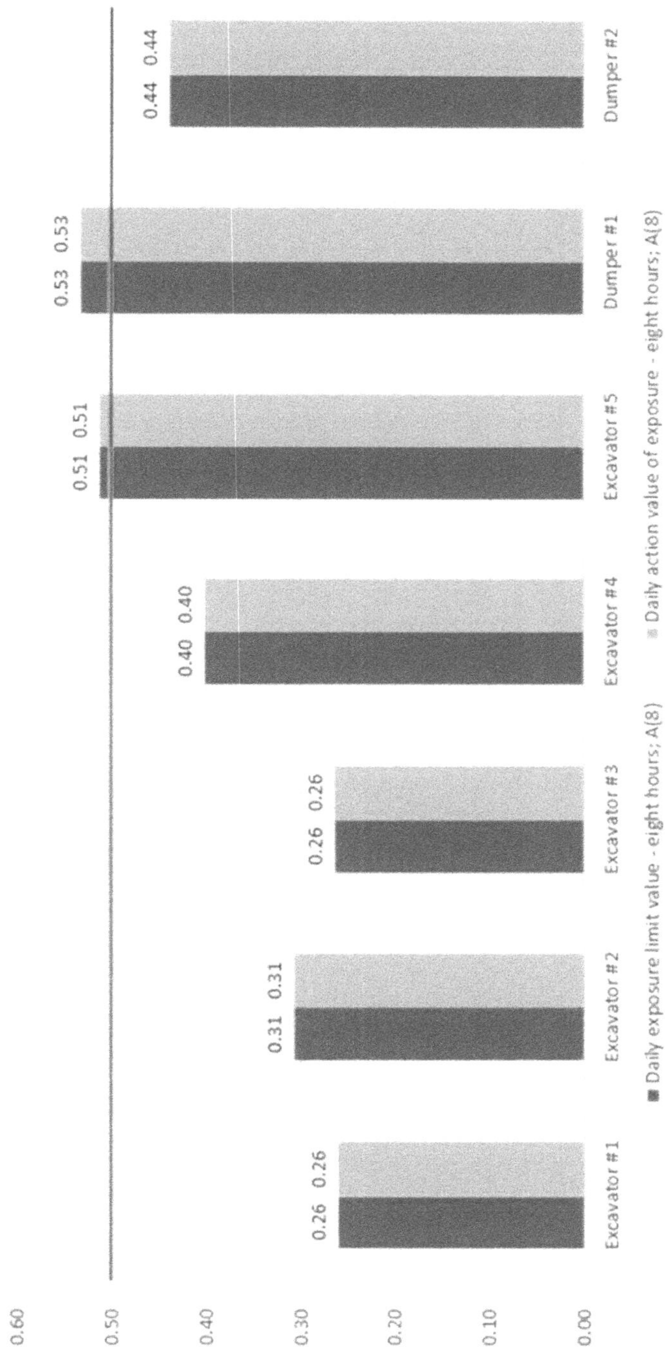

FIGURE 6.5 Recorded values of daily exposure limit value and exposure action value at the observed workplaces.

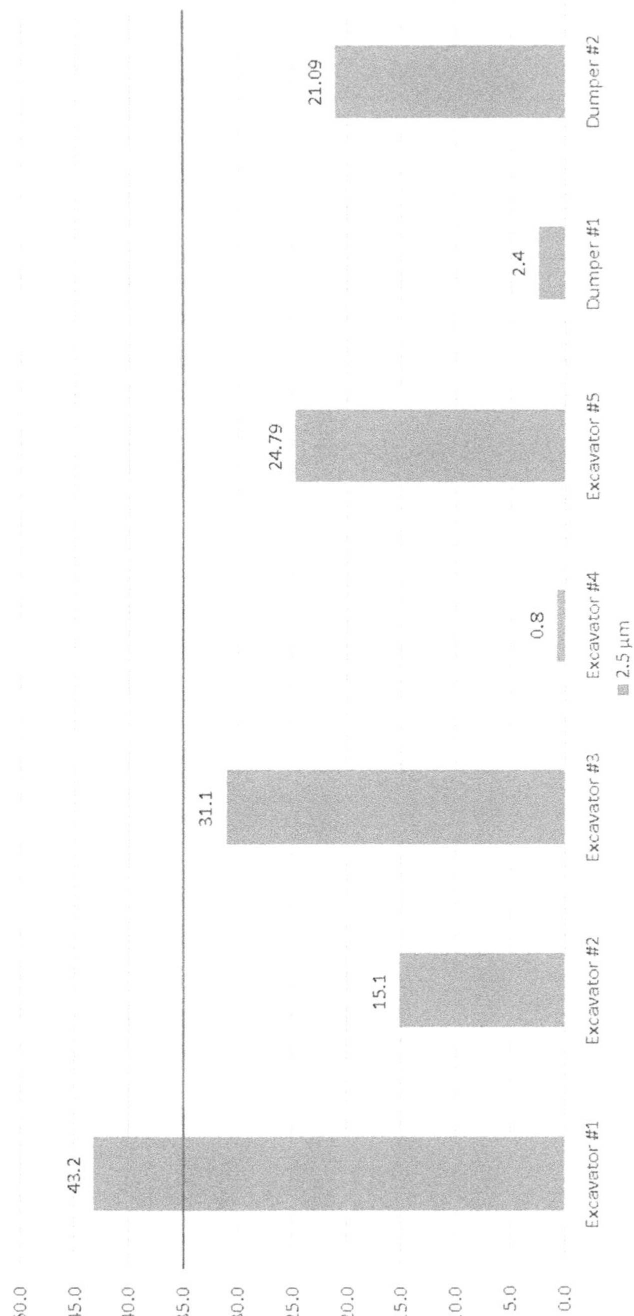

FIGURE 6.6 Recorded values of $PM_{2.5}$ concentration at the observed workplaces.

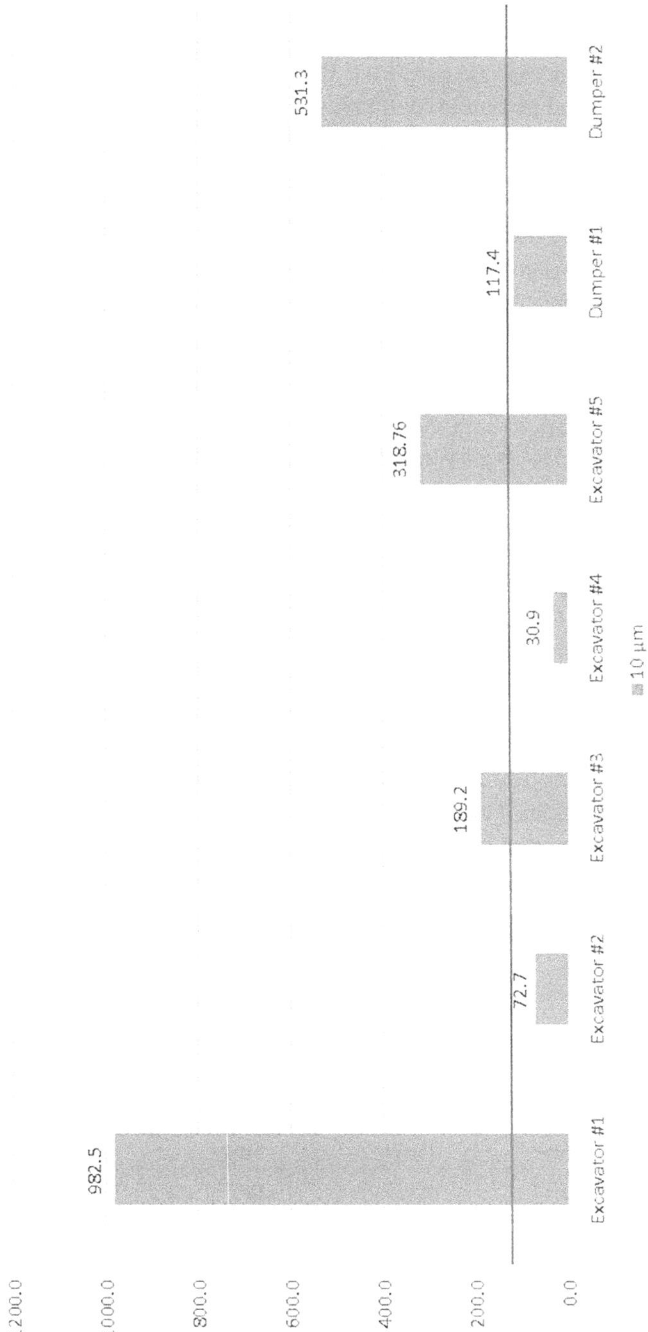

FIGURE 6.7 Recorded values of PM_{10} at the observed workplaces.

that we have recorded the exceeding of the permitted limit value in several observed mining machines (Excavator #1, #3, #5, Dumper #2).

Figure 6.8 presents the results of external exhaust gas measurements. The measurement of these quantities was realized by attaching the Testo measurement instrument to the external exhaust pipe (Figure 6.1). Directive 9768EC Measures against the emission of gaseous and particulate pollutants from internal combustion engines to be installed in non-road mobile machinery are used to estimate the emission of external gases. Stages I–V are regulations related to the emission of external gases depending on the year of production of the mining machine in question. Given that the observed mining machines are of different power and years of production, Table 6.6 gives the standard that applies to each machine individually.

Excavator #1 belongs to category R and is subject to the limit values of CO 5.0 g/kWh and NO_x 0.4 g/kWh. Excavators #2 and #5 belong to the V/C6 category and are subject to the limit values of CO 3.5 g/kWh and NO_x 0.4 g/kWh. Excavators #3 and #4 belong to category Q and the limit values of CO 3.5 g/kWh and NO_x 0.4 g/kWh apply to them. Dumpers #1 and #2 belong to category H, with the limit values of CO 3.5 g/kWh and NOx 4.0 g/kWh apply to them. Exceeding the limit value for the measured amount of carbon monoxide (CO) was recorded at Excavator #1, Excavator #3, Dumper #1, and Dumper #2, with extremely high emission measured at Excavator #1. Exceeding the limit value of NO_x was recorded in all observed machines except for Dumper #2. Directive 9768EC does not cover the upper limit for SO_2, but other regulative such as 40CFR 49.129 define rules for limiting emissions of sulfur dioxide. According to this regulative SO_2 emissions from a combustion source stack must not exceed an average of 500 parts per million by volume, on a dry basis and corrected to 7% oxygen, during any three-hour period. The specified concentrations of SO_2 from the external exhaust pipes of the observed mining machinery were not exceeded.

Figure 6.9 shows the results of measuring carbon dioxide – CO_2 concentration inside the operator's cabin. The permitted limit value for CO_2 at the operator's workplace according to Directive 2006/15/EC (second list) is 9,000 mg/m^3 or 5,000 ppm. As can be seen from the picture, none of the observed work machines exceeded the permitted values of CO_2.

TABLE 6.6
Observed Mining Machinery Characteristics – Age, Power, Stage, and Category

Type	Year	Power	Stage	Category
Excavator #1	2018	KW123	IV	R
Excavator #2	2024	KW180	V	Category V/C 6
Excavator #3	2017	KW370	IV	Q
Excavator #4	2017	KW370	IV	Q
Excavator #5	2023	KW220	V	Category V/C 6
Dumper #1	2008	KW250	III	H
Dumper #2	2010	KW298	IIIb	H

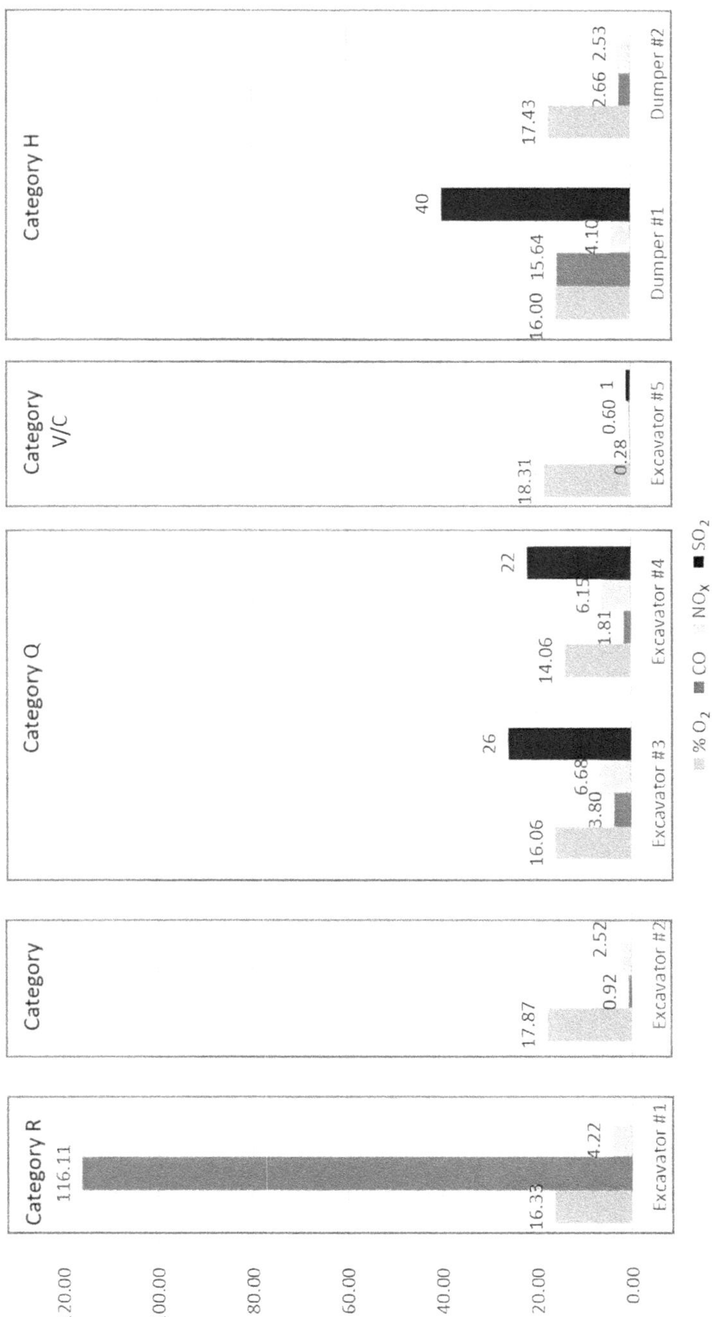

FIGURE 6.8 Recorded values of external exhaust gases at the observed workplaces.

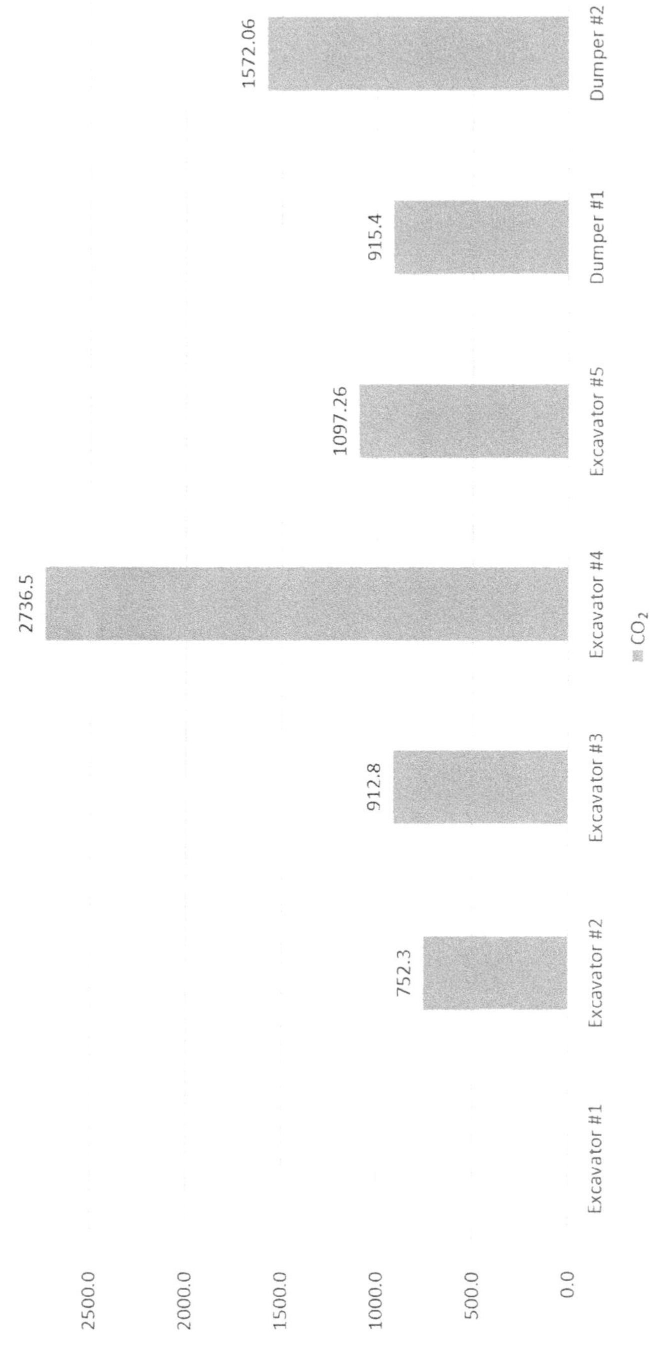

FIGURE 6.9 Recorded values of CO_2 inside the operator's cabin at the observed workplaces.

Figure 6.10 shows the results of measuring sulfur dioxide (SO_2) concentration inside the operator's cabin. The permitted limit value for SO_2 at the operator's workplace according to Directive 2006/15/EC (fourth list) is 1.3 mg/m^3 or 0.5 ppm. In Figure 6.10, the limit value of 1.3 mg/m^3 is marked with a horizontal line, and the recording results indicate that this limit value was exceeded at Excavator #2. Figure 6.10 also shows the results of measuring nitrogen dioxide – NO_2 inside the cabin of the observed mining machines. The permitted limit value of NO_2 at the operator's workplace according to Directive 2017/164/EU (fourth list) is 0.96 mg/m^3, i.e., 0.5 ppm. In Figure 6.10, the limit value of 0.96 mg/m^3 is marked with a horizontal line, and it can be seen that the exceeding of this value is also present only with excavator #2.

6.5 CONCLUSION

Improving conditions that can be of influence to workers' and residents' health in the vicinity of mining sites is imperative. This chapter's goal was to examine potential occupational hazards and ergonomic concerns related to operating dumpers and excavators. In total, 120 operators – 73 excavator and 47 dumper operators – participated in the survey, while for 50 operators, anthropometric measurements were collected. Subsequently, information on environmental quality parameters, including noise, indoor air quality, gaseous and particle emissions, and levels of vibration that are acceptable to humans, was measured on site on two dumpers and five excavators and were compared to regulatory documents. The findings demonstrate that, even with machines from more recent manufacturing dates (2023, 2024), there are exceedances of the permitted values of specific parameters of working conditions.

Based on the results of the survey, excavator operators gave more negative ratings regarding discomfort and reduced comfort in the area of the arms, legs, trunk, head, neck, and back (average rating 1.43–2.1) compared to dump truck drivers, whose average ratings on the same questions are in the range of 3–3.4 (mean), which indicates the ergonomic inadequacy of the excavators' seat. The results of surveys of excavators with regard to the negative impact of vibrations gave mean scores of 2.8–3.23, while dumper operators evaluated the impacts of vibrations with a mean score in the range of 2.4–3.4. The vibration monitoring with a sensor measurement instrument did not record exceeding the daily limit, but it did record exceeding of action daily eight-hour vibration limit at two workplaces (Excavator #5 and Dumper #1). The surveyed excavator operators gave an average score of 2.73–3.13 for disturbance in terms of noise intensity, while the dumper operators rated it as 2.8–3.48, and the measurement devices did not record exceeding the permitted noise level. Excavators assessed the unfavorable temperature in the cabin with an average score of 2.93, while dumper operators rated the same with 3.6, and indeed the measurement devices showed the presence of exceeding the air temperature inside the cabin in both observed dumpers and three of the observed five excavators.

Excavator operators gave average marks for the unfavorable impact of dust in the cabin air in range 2.87–3.03, while dumper operators gave marks in range 2.6–3. Measurement devices for recording suspended particles with a size of 10 μm

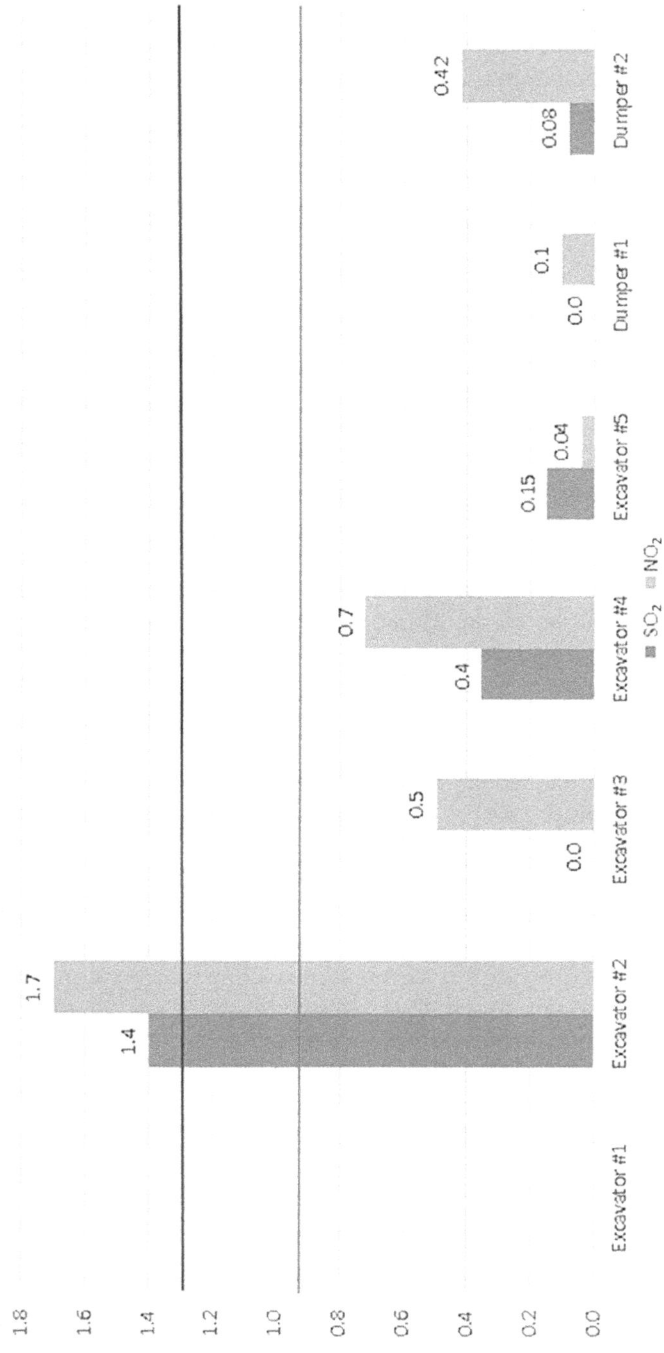

FIGURE 6.10 Recorded values of SO_2 and NO_2 inside the operator's cabin at the observed workplaces.

exceeded in three out of five recorded excavators and in one of the two dump trucks. Surveyed excavator operators rated the level of nuisance of exhaust gases in the cabin with 2.7, while dumpers gave an average rating of 3.4 for the same. While the measurement devices did not record an exceedance of CO_2, NO_2, and SO_2 levels, the level of CO was exceeded at two machines, SO_2 at one machine, as well as PM_{10} that was exceeded in three out of five observed excavator units and at one dumper.

Excavator operators assessed the level of nuisance of exhaust gases in the vicinity of the machine with an average rating of 2.83, while dumpers rated the same with 3, while the measurement instruments showed the exceeding permissible value of NOx in all observed excavators and in one of the two observed dumpers.

The subjective feeling and observations from the operator's point of view are present in terms of reduced reality for the existence of harmful gases among all the surveyed operators of mining machinery. This conclusion follows from the fact that the measurement instruments for monitoring the emission of harmful gases indicated the existence of exceeding the harmful gases (NO_x, CO, SO_2) and suspended particles (PM_{10}). The research suggests that although the operators of mining machinery are not completely aware of the presence of harmful gases, the wearing of protective equipment (masks, helmets, vests, etc.) and prevention measures should be applied more intensively, because there is a relatively high mean score on absence from work due to illness because of inadequate working conditions. The rating in terms of absence from work due to inadequate working conditions for excavators is 2.97, while for dumpers it is 3.2. Within the context of this research, and based on the obtained experimental data, the most common risks for excavator and dumper machinery operators at work evidently are related to the potential presence of human vibrations, temperature, $PM_{2.5}$, PM_{10}, NO_2, and SO_2 inside the cabin, and CO and NO_x from outside in relation to the limit values specified by the regulations. All excavators and dumpers show no excess of noise. However, only Excavator #5 and Dumper #1 exhibit an excess of human vibrations. Excavators #1, #2, and #3 show excess of temperature. Furthermore, only Excavator #1 has an excess of $PM_{2.5}$ prescribed value. Excavators #1, #3, and #5, as well as Dumper #2, show excess of PM_{10}, indicating particulate matter presence over prescribed boundaries. No excavators or dumpers show CO_2 levels over boundaries. Excavator #2 shows exceeded value of NO_2. Similar to NO_2, only Excavator #2 has shown an exceeded value of SO_2 reading. Excavators #1, #3, and Dumper #1 and #2 have shown an exceeded value for CO from outside, indicating its presence over limits. All machines except Dumper #2 show exceeded values of NO_x from outside, indicating widespread presence. Also, it is evident that operators' subjective feelings expressed by their attitude toward the questionnaire are not completely comparable to data obtained from measurement instruments. For example, statistically significant differences in subjective feelings of excavator and dumper operators have been found for certain questions about seat ergonomic characteristics including armrests, location and usage of controls, levers and pedals, size of the cabin and entrance possibility, discomfort/pain in head and trunk, visibility and general view inside and outside the cabin, control of temperature and humidity, and the atmosphere of cooperation and togetherness among the operators, whereas only temperature is evidenced also by measurement instruments. Given the results, it would be advantageous to prototype a cutting-edge multisensorial support system that will constantly monitor possible occupational risks at the level of excavators' and dumpers' workplaces. Collected anthropometric measurements and calculated percentile values could be used when designing ergonomically designed and placed multisensorial support systems. Additionally, more data gathering is encouraged.

ACKNOWLEDGMENT

This research was supported by the Science Fund of the Republic of Serbia (#GRANT No. 5151, Support Systems for Smart, Ergonomic and Sustainable Mining Machinery Workplaces – SmartMiner) and the Ministry of Science, Technological Development and Innovations (Contract no. 451-03-65/2024-03/200105 from February 5, 2024 and 337-00-216/2023-05/244 from June 17, 2024). The authors also thank the respondents who filled out the questionnaires for their kind cooperation.

REFERENCES

1. Ismail, S. N., Ramli, A., & Aziz, H. A. (2021). Research trends in mining accidents study: A systematic literature review. *Safety science*, 143, 105438.
2. Dong, L., Tong, X., Li, X., Zhou, J., Wang, S., & Liu, B. (2019). Some developments and new insights of environmental problems and deep mining strategy for cleaner production in mines. *Journal of Cleaner Production*, 210, 1562–1578.
3. Spasojević Brkić, V., Mihajlović, I., Nikolić, Đ., & Brkić, A. (2024). Smart, ergonomic and sustainable mining machinery workplaces: An overview of the Smartminer project. *IMCSM Proceedings – International May Conference on Strategic Management – IMCSM24*, May 31, 2024, Bor, 20(1), 49–59. doi: 10.5937/IMCSM24005S.
4. Tadic, D., Djapan, M., Misita, M., Stefanovic, M., & Milanovic, D. D. (2012). A fuzzy model for assessing risk of occupational safety in the processing industry. *International Journal of Occupational Safety and Ergonomics*, 18(2), 115–126.
5. Spasojević Brkić, V., Perišić, M., Veljković, Z., Misita, M., & Damjanović, M. (2022). Pareto analysis in bulldozer's failures and stoppages risk management. *Industrija*, 50(2), 37–49.
6. Williams, J. P. (2012). Global trends and tribulations in mining regulation. *Journal of Energy & Natural Resources Law*, 30(4), 391–422.
7. Du, W., & Li, M. (2020). Assessing the impact of environmental regulation on pollution abatement and collaborative emissions reduction: Micro-evidence from Chinese industrial enterprises. *Environmental Impact Assessment Review*, 82, 106382.
8. Kirin, S., Sedmak, A., Li, W., Brzaković, M., Miljanović, I., Petrović, A., & Sedmak, S. (2021). Human factor risk management procedures applied in the case of open pit mine. *Engineering Failure Analysis*, 126, 105456.
9. Spasojević Brkić, V., Misita, M., Perišić, M., Brkić, A., & Veljković, Z. (2022). Validating measurement structure of checklist for evaluating ergonomics risks in heavy mobile machinery cabs. *Mathematics*, 11(1), 23.
10. Misita, M., Brkić, A., Mihajlović, I., Đurić, G., Stanojević, N., Bugarić, U., & Spasojević Brkić, V. (2024). Decision support system for mining machinery risk mitigation driven by ergonomics and contextual theory. *Applied Sciences*, 14(15), 6413.

11. Milijic, N., Mihajlovic, I., Strbac, N., & Zivkovic, Z. (2013). Developing a questionnaire for measuring safety climate in the workplace in Serbia. *International Journal of Occupational Safety and Ergonomics*, 19(4), 631–645.
12. Papic, N., Misita, M., Spasojevic Brkic, V., Perišić, M., & Janev, N. (2024). Dumper operator's workplace risks: Preliminary study. *Human Factors and Systems Interaction, AHFE (2024) International Conference*, 154, 49–58. doi: 10.54941/ahfe1005352.
13. Akyeampong, J., Udoka, S., Caruso, G., & Bordegoni, M. (2014). Evaluation of hydraulic excavator human–machine interface concepts using NASA TLX. *International Journal of Industrial Ergonomics*, 44(3), 374–382.
14. Souza, J., Castelo Branco, J., & Baptista, J. S. (2023). Work accidents related to heavy equipment in the open pit extractive industry: A systematic review. *Occupational and Environmental Safety and Health*, V, 783–791.
15. Perišić, M., Spasojevic Brkic, V., Mihajlovic, I., Brkic, A., & Janev, N. (2024). Excavators' cabins ergonomic design influential factors modelling: Preliminary study. *Human Factors and Systems Interaction, AHFE (2024) International Conference*, 154, 39–48. doi: 10.54941/ahfe1005351.
16. Spasojević Brkić, V., Klarin, M., & Brkić, A. (2015). Ergonomic design of crane cabin interior: The path to improved safety. *Safety Science*, 73, 43–51. doi: 10.1016/j.ssci.2014.11.015.
17. Eger, T., Stevenson, J. M., Grenier, S., Boileau, P. É., & Smets, M. P. (2011). Influence of vehicle size, haulage capacity and ride control on vibration exposure and predicted health risks for LHD vehicle operators. *Journal of Low Frequency Noise, Vibration and Active Control*, 30(1), 45–62.
18. Boudreau-Trudel, B., Nadeau, S., Zaras, K., & Deschamps, I. (2014). Introduction of innovative equipment in mining: Impact on occupational health and safety. *Open Journal of Safety Science and Technology*, 4(1), 49.
19. Mandal, B., & Srivastava, A. K. (2010). Musculoskeletal disorders in dumper operators exposed to whole body vibration at Indian mines. *International Journal of Mining, Reclamation and Environment*, 24(3), 233–243. doi: 10.1080/17480930903526227.
20. Upadhyay, R., Jaiswal, V., Bhattacherjee, A., & Patra, A. K. (2022). Role of whole-body vibration exposure and posture of dumper operators in musculoskeletal disorders: A case study in metalliferous mines. *International Journal of Occupational Safety and Ergonomics*, 28(3), 1711–1721. doi: 10.1080/10803548.2021.1932111.
21. Moon, K. B., Park, M. Y., Hwang, S. H., Lee, E., Hong, K., Lim, H. J., & Lee, K. M. (2015). Subjective symptoms of musculoskeletal disorders among excavator drivers. *Journal of Korean Society of Occupational and Environmental Hygiene*, 25(1), 13–19.
22. Ramar, K., Kumaraswamidhas, L. A., Balaji, P. S., & Agasthian, A. (2023). Whole body vibration impact assessment on dumper operator using computational learning technique. *International Journal of Precision Engineering and Manufacturing*, 24(2), 219–238. doi: 10.1007/s12541-022-00732-0.
23. Kittusamy, N. K. (2003). A checklist for evaluating cab design of construction equipment. *Applied Occupational and Environmental Hygiene*, 18(10), 721–723.
24. Kittusamy, N. K., & Buchholz, B. (2000). Assessment of ergonomic exposures among operators of construction equipment. *Proceedings of the Human Factors and Ergonomics Society Annual Meeting*, 44(29), 173–176.
25. Kittusamy, N. K., & Buchholz, B. (2004). Whole-body vibration and postural stress among operators of construction equipment: A literature review. *Journal of Safety Research*, 35(3), 255–261.

26. Zunjic, A., Spasojevic Brkic, V., Klarin, M., Brkic, A., & Krstic, D. (2015). Anthropometric assessment of crane cabins and recommendations for design: A case study. *Work*, 52(1), 185–194.
27. Jeripotula, S. K. K., Mangalpady, A., & Mandela, G. R. R. (2020). Assessment of exposure to whole-body vibration of dozer operators based on postural variability. *Mining, Metallurgy & Exploration*, 813–820.
28. Lin, C. C. (2011). Ergonomic assessment of excavator seat. *International Journal of Applied Science and Engineering*, 9(2), 99–109.
29. Joy, G. J., & Middendorf, P. J. (2007). Noise exposure and hearing conservation in US coal mines – a surveillance report. *Journal of Occupational and Environmental Hygiene*, 4(1), 26–35.
30. Kurtz, L. A., Vi, P., & Verma, D. K. (2012). Case study: Occupational exposures to hand-arm vibration, whole-body vibration, and noise among crane operators in construction: A pilot study. *Journal of Occupational and Environmental Hygiene*, 9(6), D117–D122.
31. Sun, K., & Amanda S. A. (2018). Evaluating hearing loss risks in the mining industry through MSHA citations. *Journal of Occupational and Environmental Hygiene*, 15(3), 246–262.
32. Lutz, E. A., Reed, R. J., Lee, V. S., & Burgess, J. L. (2015). Occupational exposures to emissions from combustion of diesel and alternative fuels in underground mining – a simulated pilot study. *Journal of Occupational and Environmental Hygiene*, 12(3), 18–25.
33. Djordjevic, P., Nikolic, D., Jovanovic, I., Mihajlovic, I., Savic, M., & Zivkovic, Z. (2013). Episodes of extremely high concentrations of SO_2 and particulate matter in the urban environment of Bor, Serbia. *Environmental Research*, 126, 204–207.
34. Alsharif, A. M., Spasojević Brkić, V., Misita, M., Mihajlović, I., Brkić, A., Papić, N., & Perišić, M. (2024). The safety climate, hierarchical levels and resilience assessment in transport and mining companies. *Civil Engineering and Environmental Systems*, 44–68.
35. Hoffmann, E. R., & Chan, A. H. (2018). Review of compatibility and selection of multiple lever controls used in heavy machinery. *International Journal of Industrial Ergonomics*, 65, 93–102.
36. Singh, K. J., Palei, S. K., & Karmakar, N. C. (2024). Role of contributing factors on health risks of whole-body vibration exposure of heavy equipment and vehicle operators: A critical review. *Journal of Vibration and Control*, 30(11–12), 2338–2355.
37. Caffaro, F., Cremasco, M. M., Preti, C., & Cavallo, E. (2016). Ergonomic analysis of the effects of a telehandler's active suspended cab on whole body vibration level and operator comfort. *International Journal of Industrial Ergonomics*, 53, 19–26.
38. Chi, F., Zhou, J., Zhang, Q., Wang, Y., & Huang, P. (2017). Avoiding the health hazard of people from construction vehicles: A strategy for controlling the vibration of a wheel loader. *International Journal of Environmental Research and Public Health*, 14(3), 275.
39. Hoogendoorn, W. E., van Poppel, M. N., Bongers, P. M., Koes, B. W., & Bouter, L. M. (2000). Systematic review of psychosocial factors at work and private life as risk factors for back pain. *Spine*, 25(16), 2114–2125
40. Spasojević Brkić, V., Mihajlović, I., Brkić, A., Misita, M., & Perišić, M. (2024). Transport and mining machinery cabins' ergonomic evaluation as a path to its redesign. *International Journal of Occupational Safety and Ergonomics*, 30(3), 896–906.
41. Munro, D. M., Govers, M. E., & Oliver, M. L. (2021). Physical demands of overhead crane operation. *International Journal of Industrial Ergonomics*, 86, 103200.

42. ISO 9612:2009 Acoustics – Determination of occupational noise exposure – Engineering method, International Organisation for Standardisation, Geneva, Switzerland.
43. ISO 5128:2023. Acoustics – Measurement of noise inside motor vehicles, International Organisation for Standardisation, Geneva, Switzerland.
44. ISO 11201:2010. Acoustics – Noise emitted by machinery and equipment – Determination of emission sound pressure levels at the work station and at other specified positions in an essentially free field over a reflection plane with negligible environmental corrections, International Organisation for Standardisation, Geneva, Switzerland.
45. Directive 97/68EC Measures against the emission of gaseous and particulate pollutants from internal combustion engines to be installed in non-road mobile machinery 2010, European Union: Brussels, Belgium.
46. Directive 981053 Environmental protection – The Non-Road Mobile Machinery (Emission of Gaseous and Particulate Pollutants) Regulations 1999, European Union: Brussels, Belgium.
47. Directive 2008/50/EC of the European Parliament and of the Council of 21 May 2008 on ambient air quality and cleaner air for Europe. PM2.5 objectives targeting exposure population to fine particles, Official Journal of the European Union. European Union: Brussels, Belgium.
48. Regulation (EU) 2019/1242 setting CO_2 emission performance standards for new heavy-duty vehicles and amending Regulations (EC) No 5952009 and (EU) 2018956 of the Europa 2019, 96/53/EC: 2019/1242. *Official Journal of the European Union*, 50, 202–240.
49. ISO 2631-1:1997. Mechanical vibration and shock – Evaluation of human exposure to whole-body vibration – Part 1: General requirements, International Organisation for Standardisation, Geneva, Switzerland.
50. ISO 2631- 5:2004. Mechanical vibration and shock – Evaluation of human exposure to whole-body vibration – Part 5: Method for evaluation of vibration containing multiple shocks, International Organisation for Standardisation, Geneva, Switzerland.
51. ISO-2631:1978 Guide for the evaluation of human exposure to whole-body, International Organisation for Standardisation, Geneva, Switzerland.
52. ISO-TR-25398:2006. Earth-moving machinery – Guidelines for assessment of exposure to whole-body vibration of ride-on machines – Use of harmonized data measured by international institutes, organizations and manufacturers, International Organisation for Standardisation, Geneva, Switzerland.

7 Muscle Health and Function as a New Vital Sign of COVID-19 Mortality

Atabak Allafasghari, Sajjad Rostamzadeh, Fatemeh Chaharaghran, Amin Allafasghari and Alireza Abouhossein

7.1 INTRODUCTION

In the domain of occupational performance, various factors influence an individual's ability to perform daily tasks efficiently and safely. Among these, handgrip strength (HGS) and muscle endurance time (MET) are widely recognized as non-invasive biomarkers for assessing overall muscle health. These parameters are particularly valuable in predicting outcomes related to mobility, disease progression, and mortality (1–9). Evaluating HGS and MET enables healthcare professionals to identify challenges and limitations that may hinder occupational performance, facilitating targeted interventions and tailored support across diverse settings (10, 11).

HGS serves as a key indicator of muscular strength and general health, significantly influencing an individual's ability to perform tasks that require gripping, lifting, or manipulating objects (1, 12). Research indicates that reduced HGS is associated with increased risks of falls, disability, and decreased independence in daily activities (13, 14). Similarly, MET, defined as the duration an individual can sustain a specific muscle contraction or activity, is critical for maintaining consistent performance throughout the day (15). Adequate endurance reduces the likelihood of fatigue-related accidents or injuries, ensuring optimal occupational functionality (16, 17). Together, HGS and MET contribute to precise motor control, enabling individuals to perform intricate tasks, maintain balance, and adapt to changing environmental demands.

The COVID-19 pandemic has underscored the importance of identifying factors that influence disease severity and patient outcomes. HGS and MET have garnered attention as simple yet effective measures of muscular strength and health (8, 10, 11). Numerous studies highlight the association between low grip strength (LGS) and adverse health outcomes, including frailty, sarcopenia, and prolonged hospital stays (18, 19). For instance, Kara et al. reported that COVID-19 patients with low HGS were three times more likely to develop pneumonia compared to those with higher HGS (20). Similarly, Rostamzadeh et al. demonstrated that HGS, alongside common clinical markers such as chronic obstructive pulmonary disease (COPD),

DOI: 10.1201/9781003672036-7

C-reactive protein (CRP), and oxygen saturation (SaO_2), can serve as a cost-effective prognostic tool for predicting mortality risk in COVID-19 patients (11, 21). As our understanding of COVID-19 evolves, researchers and healthcare professionals are increasingly examining the relationship between HGS, MET, and the risk of severe illness, complications, and mortality. Studies by Pijls et al. and Guan et al. emphasize the relationship between age, underlying health conditions, and increased risks of severe outcomes in COVID-19 patients (22, 23). The severe acute respiratory syndrome coronavirus 2 (SARS-CoV-2), responsible for COVID-19, triggers complex pathophysiological processes that exacerbate disease severity. Elevated levels of inflammatory mediators such as TNF-α, IL-1β, IL-6, and CRP have been observed in severe cases, contributing to cytokine storms, organ damage, and heightened mortality risk (24–28). Furthermore, weakened muscle function, particularly respiratory muscle dysfunction, in individuals with acute or inflammatory lung diseases can lead to prolonged ICU stays and increased mortality (18, 29).

Given the global challenges posed by the COVID-19 pandemic and its socioeconomic impact (30, 31), understanding factors that contribute to disease severity and mortality is imperative for effective risk management. However, predictive models of COVID-19 outcomes are often hindered by strong multicollinearity among explanatory variables, incorrect data segmentation, and the inclusion of irrelevant features, reducing the reliability of such studies. To our knowledge, this research is among the few studies in clinical medicine and physical ergonomics to objectively examine the role of HGS and MET – key measures of physical frailty and muscle function – in predicting COVID-19 mortality among ICU patients. This investigation aims to address existing gaps by integrating robust clinical and physical assessments, providing valuable insights into the relationship between muscle health and COVID-19 outcomes.

7.2 MATERIAL AND METHODS

In this retrospective study, electronic medical records were utilized to identify the association between HGS and MET in predicting COVID-19 mortality among patients admitted to the ICU. The inclusion criteria for patient selection were:

- Patients aged 27–90 years;
- Laboratory-confirmed SARS-CoV-2 infection;
- Underwent concomitant systematic variant screening;
- No history of fracture, deformity, or surgery in the upper extremities in the past year;
- No presence of neuromuscular or rheumatic diseases affecting HGS measurements;
- Admission to the ICU between April and September 2021;

Exclusion criteria included patients with appendicular skeletal muscle mass index scores below 7.0 kg/m² who were diagnosed with sarcopenia before COVID-19. According to World Health Organization (WHO) interim guidance, patients admitted to the ICU exhibited clinical signs of pneumonia, respiratory rates above 30/min, oxygen saturation (SpO_2) at or below 80%, and extensive lung involvement in computed tomography (CT) scans while on room air.

The study included a total of 2175 COVID-19 patients (1025 females and 1150 males) aged 27–90 years who were admitted to the ICU. Data collection and analysis were performed after the patients were either discharged alive or had died in the ICU. The study received approval from the Research and Ethics Committees of Shahid Beheshti University of Medical Sciences (IR.SBMU.PHNS.REC.1402.087), and written informed consent was obtained from all participating patients or their immediate family members to use the patients' health records.

By selecting patients with these specific criteria, the researchers aimed to investigate the predictive value of HGS and MET in COVID-19 mortality in the ICU, potentially identifying high-risk patients and informing targeted interventions and treatment strategies (11, 21).

Various data points were collected from the medical records of patients aged 27–90 years admitted to the ICU with COVID-19. These data included:

- Demographic information: age, gender;
- Smoking status;
- Comorbidities: presence of underlying health conditions such as hypertension, overweight, diabetes mellitus (DM), hypothyroidism, cardiovascular disease (CVD), bronchial asthma, COPD, and cancer;
- COVID-19-related symptoms: cough, fever, myalgia, dyspnea, anosmia/ageusia, diarrhea, and asymptomatic cases;
- Clinical markers: CRP, ferritin, D-dimer, SaO_2, white blood cell (WBC), thrombocyte, lymphocyte, neutrophil, and hemoglobin levels;
- Physical measurements: stature (cm) using a Holtain Harpenden stadiometer, and body mass (kg) using a digital balance, with patients wearing light clothing but no shoes;
- Body Mass Index (BMI): calculated as kg/m^2 and classified into four groups – underweight (BMI <18.5), healthy weight (18.5–24.9), overweight (25.0–29.9), and obese (BMI ≥30) (32).

The comprehensive data collection aimed to provide an in-depth understanding of patients' health status, encompassing demographic characteristics, pre-existing conditions, COVID-19-related symptoms, and clinical markers. This dataset facilitated the analysis of the relationship between HGS and MET in predicting COVID-19 mortality among ICU patients, while also enabling the identification of potential risk factors contributing to adverse outcomes.

CT scans are a crucial diagnostic tool for evaluating lung involvement and disease progression in COVID-19 patients. A comprehensive assessment of CT findings is essential to determine the extent and severity of lung abnormalities. This study describes the methodology and scoring system used to analyze patients' CT scans and quantify radiological findings in COVID-19 patients.

CT imaging was performed with patients positioned supine at the end of inspiration using a GE Bright speed 16-Slice CT Scanner (GE Healthcare, Chicago, Illinois, United States) (33). The scans were analyzed by an experienced radiologist, focusing on lung infiltration patterns, including ground-glass opacities (GGO), GGO with consolidation, consolidation, and other findings such as linear opacities, traction bronchiectasis, cysts, and reticular opacities.

Both initial chest CT scans and follow-up scans (if available) were evaluated using this methodology (34). The anatomical distribution of infiltrations was categorized according to predominant imaging findings in five lung zones (three lobes in the right upper, middle, and lower, and two lobes in the left upper and lower). A semi-quantitative CT severity scoring system was applied to quantify radiological findings in each lobe. The scoring system assigned scores as follows: 0 (no involvement), 1 (<5%), 2 (5–25%), 3 (26–50%), 4 (51–75%), and 5 (>75%). The total CT score was calculated by summing the scores for all lobes, yielding a range of 0 to 25 (35). This standardized methodology enabled an objective assessment of lung abnormalities, providing valuable insights into disease severity, progression, treatment response, and potential long-term implications of COVID-19-associated lung pathology.

Maximum Voluntary Contraction (MVC) and MET in COVID-19 patients using the JAMAR® hydraulic dynamometer, as these parameters provide valuable insights into muscle function and overall physical performance. The protocol adhered to established guidelines for grip strength assessment and incorporated gender-specific cut-offs for LGS. To determine hand dominance, participants were asked to identify the hand used for most routine tasks. They were seated with their feet flat on the floor, arms relaxed at their sides in a neutral position, elbows flexed at 90°, and forearms and wrists positioned neutrally (0–15° extension and 0–15° ulnar deviation) (36). For MVC assessment, participants were instructed to exert maximal voluntary contraction on the dynamometer for 3 seconds using their dominant hand. Three MVC trials were recorded, with a 1-minute rest period between each exertion, and the average value was used for further analysis. LGS was determined using gender-specific thresholds derived from peak mean values of healthy adults: <26 kg for males and <14 kg for females (37).

MET was assessed in an additional trial. Patients were asked to maintain their grip, and the time (in seconds) was recorded until their grip strength dropped to 50% of maximum level (38).

7.2.1 Statistical Analysis of the Data

Statistical analyses were conducted using SPSS 23 (IBM Corporation, New York, NY, United States). The Kolmogorov–Smirnov test was employed to assess the normality of the data. Outliers were identified using Grubb's test, which detects extreme values based on their deviation from the sample mean relative to the standard deviation (SD) (39). The reliability of grip strength and muscle endurance tests in COVID-19 patients was evaluated using the Intra-class Correlation Coefficient (ICC) for relative reliability and the standard error of measurement (SEM) for absolute reliability.

Descriptive statistics, including means, SDs, minimum-maximum values, and percentages, were calculated for clinical and demographic characteristics as well as chest CT scores. An independent samples *t*-test was used to compare clinical characteristics between male and female patients. For comparing numerical variables across groups, one-way ANOVA was conducted, followed by the Tukey post-hoc test for pairwise comparisons. Analysis of covariance (ANCOVA) was used to determine the effects of age, gender, and the interaction between age and gender on HGS in survived and deceased patients. Categorical variables were analyzed using Chi-square test or Fisher's exact test, as appropriate. Sensitivity analyses were performed on patient subgroups to identify potential sources of heterogeneity and assess the robustness of the findings. The consistency between sensitivity analyses and primary results suggested that the assumptions made for the main analysis had minimal impact on the overall conclusions, thereby enhancing the reliability of the study findings. Statistical significance was defined as $p < 0.05$.

To identify potential predictors of mortality among ICU-admitted COVID-19 patients, binary logistic regression analysis was conducted using the backward likelihood ratio selection method. Patients were classified into two groups – deceased and non-deceased – and significant demographic and clinical variables were included in the predictive model. This approach aimed to identify key factors contributing to mortality in this critical population.

7.3 RESULTS

The sample consisted of 2175 COVID-19 patients aged 27–90 years, with 1150 males (52.9%) and 1025 females (47.1%). The mean age was 61.1 ± 8.6 for males and 74.4 ± 6.4 years for females. A total of 445 patients (289 males, 156 females) died, resulting in a high mortality rate of 20.5% (20.2% for males and 20.8% for females). A flowchart presenting the sampling procedure is shown in Figure 7.1. As indicated, handgrip strength was not recorded for 18,263 out of 20,438 patients due to the inclusion and exclusion criteria.

7.3.1 Clinical and Demographic Characteristics

Table 7.1 presents the clinical findings and demographic characteristics of COVID-19 patients, categorized by survival status (survived vs. deceased). To investigate the clinical causes of death, we reviewed the legal death certificates of the 445 deceased patients. The most frequent causes of death were infection-related, occurring in 254 cases (57.1%). These included sepsis, septic shock, or sepsis-related multi-organ failure in 125 cases (28.1%), bacterial infections in 49 cases (11.0%), and viral pneumonia in 17 cases (3.8%). The second most common causes of death were respiratory-related, documented in 117 cases (26.3%) and involving respiratory insufficiency, COVID-19 pneumonia, hypoxia, or acute respiratory distress syndrome (ARDS). Additional immediate causes of death were pulmonary embolism, cardiovascular failure, and cancer, accounting for 74 cases (16.6%). Overall, the most frequent

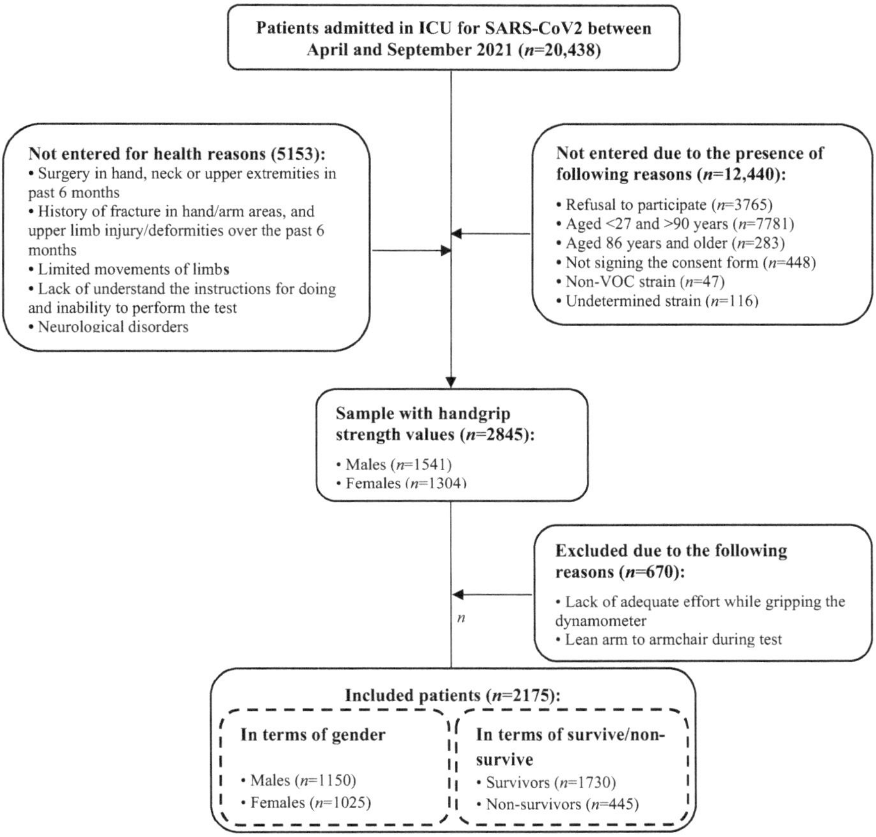

FIGURE 7.1 Sampling flowchart of the study.

symptoms and comorbidities among COVID-19 patients, along with the significance of differences (p values <0.05), are shown in bold in Table 7.1. Given that 30–50% of COVID-19 patients experienced myalgia and exhibited lower HGS, we performed sensitivity and robustness analyses using rare-events logistic regression in this subgroup. The sensitivity analysis indicated that the main results remained unchanged in the presence of COVID-19 patients with myalgia (OR = 0.67, 95% CI = 0.36–0.94, $p = 0.023$).

A significant difference was observed in mean HGS values and the prevalence of LGS between deceased and survived COVID-19 patients ($p < 0.001$). Deceased patients exhibited significantly lower HGS and a higher frequency of LGS compared to survivors. The Jamar dynamometer demonstrated high to very high test-retest reliability ($0.87 \leq ICC \leq 0.93$; $p \leq 0.001$; $n = 50$ subjects). Moreover, mean endurance time values were notably lower in deceased patients compared to survivors across both genders ($p < 0.001$).

TABLE 7.1
Clinical Biomarkers and Demographic Characteristics of the Patients Hospitalized in ICU

Characteristics	Deceased (n = 445)	Survived (n = 1730)	p value
Age (years)	68.4 ± 7.8	47.3 ± 8.9	**<0.001**
Gender			
Males	289 (66.2%)	861 (49.8%)	**<0.001**
Females	156 (33.8%)	265 (50.2%)	**<0.001**
BMI (kg/m^2)			
Underweight (<18.5)	17.6 ± 3.8	18.7 ± 4.3	0.689
Healthy weight (18.5 to 24.9)	21.9 ± 6.6	22.4 ± 7.2	0.721
Overweight (25.0 to 29.9)	29.3 ± 5.3	25.8 ± 7.7	**<0.001**
Obese (≥30)	34.1 ± 7.0	31.8 ± 8.3	**<0.001**
Smoking, *n* (%)	133 (29.9%)	294 (17.0%)	**<0.001**
Hospital stay in ICU (day)	14 (7–34)	8 (5–17)	**<0.001**
Symptoms at ICU, *n* (%)			
Cough	272 (61.1%)	689 (39.8%)	**<0.001**
Fever (>37.8 C°)	316 (71.0%)	655 (37.8%)	**<0.001**
Myalgia	147 (33.0%)	565 (32.6%)	0.117
Dyspnea	307 (69.0%)	170 (9.8%)	**<0.001**
Anosmia/ageusia	45 (10.1%)	329 (19.0%)	**0.027**
Diarrhea	30 (6.7%)	86 (5.0%)	0.069
Asymptomatic	46 (10.3%)	147 (8.5%)	**0.008**
Comorbidities, *n* (%)			
Hypertension	245 (55.0%)	570 (33.0%)	**<0.001**
Obesity	120 (26.9%)	165 (9.5%)	**<0.001**
Diabetes Mellitus	86 (19.3%)	308 (17.8%)	0.067
Bronchial asthma	112 (25.2%)	121 (7.0%)	**<0.001**
Hypothyroidism	49 (11.0%)	168 (9.7%)	0.257

Cardiovascular disease	165 (37.0%)	177 (10.2%)	**<0.001**
Cancer	24 (5.4%)	80 (4.6%)	0.448
COPD	92 (20.7%)	85 (4.9%)	**<0.001**
Laboratory, mean±SD (IQR)			
C-reactive protein (mg/l)	63.7 ± 12.5 (27.6–139.2)	30.7 ± 11.7 (21.3–77.9)	**<0.001**
Ferritin (μg/l)	254.4 ± 37.3 (157.2–728.3)	110.1 ± 25.2 (49.5–356.3)	**<0.001**
D-dimer (mg/l)	4.0 ± 1.1 (1.4–7.1)	1.4 ± 0.5 (0.9–2.5)	**<0.001**
SaO_2 (%)	68.1 ± 13.1 (49.3–83.5)	87.3 ± 7.7 (66.4–93.4)	**<0.001**
WBC ($\times 10^3$/μl)	10.8 ± 4.7 (6.2–15.1)	7.1 ± 3.4 (4.2–17.2)	**<0.001**
Thrombocyte ($\times 10^3$/μl)	165.4 ± 41.6 (110.7–248.1)	202.4 ± 47.4 (128.3–265.4)	**<0.001**
Lymphocyte ($\times 10^3$/μl)	0.7 ± 0.3 (0.5–1.6)	1.1 ± 0.2 (0.7–1.5)	0.077
Neutrophil ($\times 10^3$/μl)	6.6 ± 1.1 (3.0–8.3)	3.0 ± 1.2 (1.7–5.1)	**<0.001**
Hemoglobin (g/dl)	11.0 ± 1.7 (8.1–12.9)	12.7 ± 1.8 (9.2–14.9)	**0.037**
Handgrip strength (kg)			
Males	27.1 ± 7.8 (12.0–49.4)	36.3 ± 4.8 (9.3–57.2)	**<0.001**
Females	22.8 ± 4.6 (10.2–30.5)	28.9 ± 3.7 (8.1–42.1)	**<0.001**
Total	25.0 ± 5.8 (10.2–49.4)	32.6 ± 6.1 (8.1–57.2)	**<0.001**
Low grip strength, *n* (%)			
Males	161 (55.7%)	254 (29.5%)	**<0.001**
Females	68 (43.6%)	71 (26.8%)	**<0.001**
Handgrip endurance time (s)			
Males	15.9 ± 5.3 (7.1–31.0)	26.7 ± 6.9 (14.3–37.3)	**<0.001**
Females	8.1 ± 3.7 (3.8–18.9)	17.1 ± 4.8 (7.6–29.1)	**<0.001**

Note: BMI, body mass index; COPD, chronic obstructive pulmonary disease; IQR, interquartile range; WBC, white blood cell. Bold indicates statistically significant variables ($p < 0.05$).

These findings suggest that HGS and MET could serve as potential predictors of mortality in COVID-19 patients, highlighting the importance of assessing these parameters in clinical practices. The robust reliability of the Jamar dynamometer further supports its utility in accurately measuring grip strength and endurance time in this patient population.

7.3.2 Mean Differences in HGS and MET by Gender, Age Group, and Survived/Deceased

Table 7.2 presents the effects of age, gender, and their interaction on HGS. Age had a significant impact on HGS measurements ($F = 3892.41$ for survivors and $F = 2735.37$ for deceased patients; $p < 0.0001$). Gender also had a meaningful effect on HGS ($F = 142.27$ for survivors and $F = 133.22$ for deceased patients; $p < 0.0001$), with males exhibiting greater HGS than females across all age groups. Specifically, the average HGS of female patients was approximately 65% of that of their counterparts in the survivor group and 31.3% in the deceased group. The interaction between age and gender also had a significant effect on HGS ($F = 376.33$ for survivors and $F = 345.12$ for deceased patients; $p < 0.0001$). Additionally, the findings revealed significant effects of age ($F = 2765.24$ for survivors and $F = 2276.34$ for deceased patients; $p < 0.0001$), gender ($F = 117.56$ for survivors and $F = 101.44$ for deceased patients; $p < 0.0001$), and their interaction ($F = 328.77$ for survivors and $F = 289.65$ for deceased patients; $p < 0.0001$) on MET.

Figure 7.2 shows the trend of mean HGS for male and female patients admitted to ICU in terms of age group and survived/deceased. In general, a significant difference was observed between the HGS of deceased and survived patients in both genders. Figure 7.2a demonstrated that the HGS of male patients who died from COVID-19 was lower than their counterparts who survived (by about 11–59%, where the highest and lowest differences were observed in the male patients aged 65–69 and 35–39 years, respectively). A similar pattern can be seen for female patients in Figure 7.2b. Similar to male patients, the HGS of female patients who died from COVID-19 was lower than their counterparts who survived (by about 9–39%, where the highest and lowest differences were observed in the male patients aged 60–64 and 30–35 years, respectively).

TABLE 7.2
Univariate (Unadjusted) ANCOVA Results for Age, Gender, and Age/Gender

	$F_{1,1729}$ (Survived)			$F_{1,444}$ (Deceased)		
Variable	**Age**	**Gender**	**Age/Gender**	**Age**	**Gender**	**Age/Gender**
HGS	3892.41	142.27	376.33	2735.37	133.22	345.12
MET	2765.24	117.56	328.77	2276.34	101.44	289.65

Note: HGS, handgrip strength; MET, muscle endurance time.
All *p* values were <.0001.

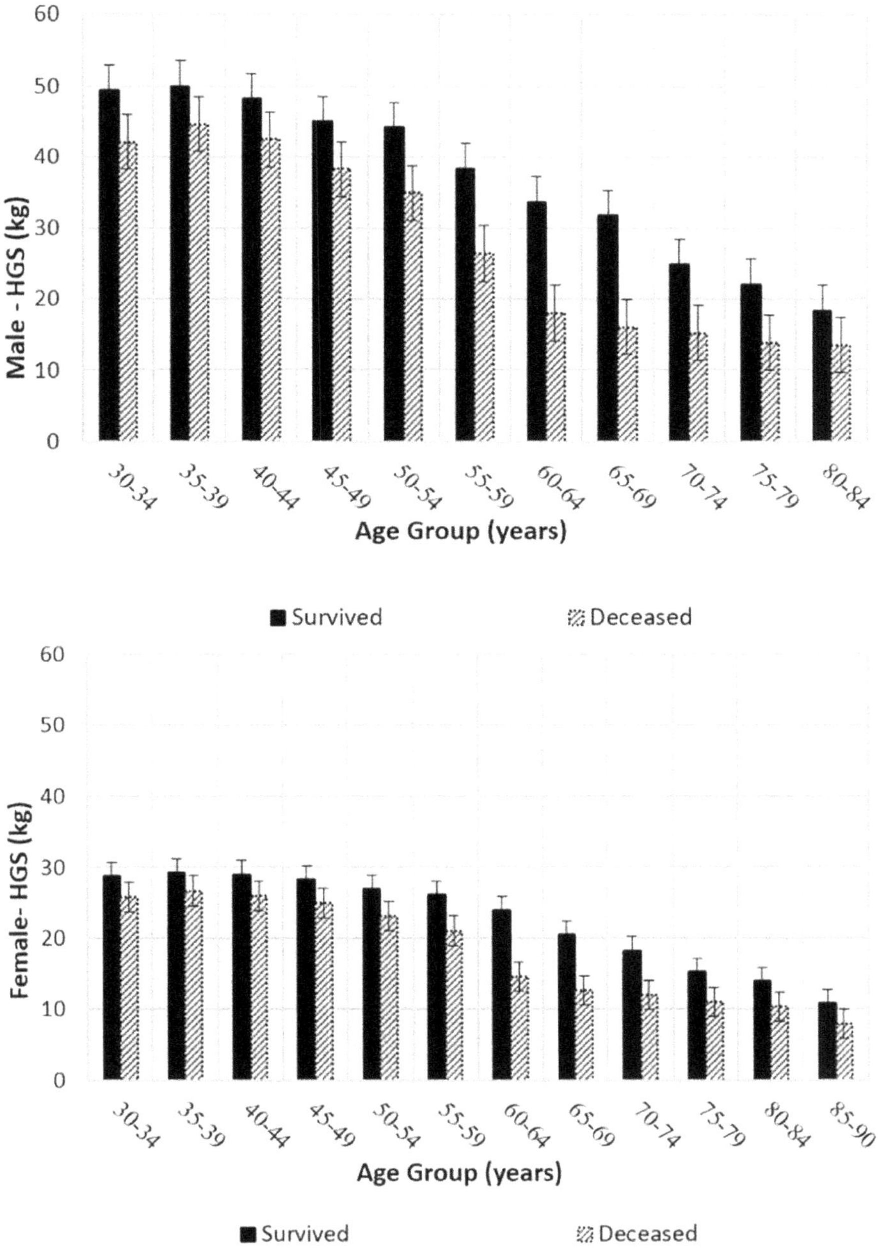

FIGURE 7.2 Mean HGS for patients admitted to ICU, in terms of age group and survived/deceased; (a) in males and (b) in females.

Note: HGS, handgrip strength.

TABLE 7.3
Binary Logistic Regression Results for the Prediction of Mortality of COVID-19 Patients Hospitalized in ICU

Dependent Variable	Independent Variable	Exp (β)	95% CI	*p*
Mortality	Age	2.447	1.274–6.117	0.002
	COPD	6.324	1.725–26.558	0.004
	SaO_2	−4.621	2.447–1.323	<0.001
	CRP	2.625	1.042–6.236	<0.001
	LGS	5.735	1.447–11.472	<0.001
	MET	3.610	1.883–8.550	<0.001

Note: β, standardized coefficients; CI, confidence interval; COPD, chronic obstructive pulmonary disease; CRP, C-reactive protein; LGS, low grip strength; MET, muscle endurance time; SaO_2, oxygen saturation.

7.3.3 Binary Logistic Regression Analyses for Predicting Mortality

Initially, 31 variables listed in Table 7.1 were mapped into the model. All factors were tested using the Variance Inflation Factor (VIF) test (40) and the correlation matrix. As a result, 19 independent variables were included in the model, which includes age, gender, BMI, smoking status, LGS, MET, hypertension, D-dimer, Ferritin, SaO_2, C-reactive protein, cardiovascular disease, COPD, bronchial asthma, thrombocyte, hemoglobin, diabetes mellitus, WBC, neutrophil. However, the step-wise analysis gave the best-fitted model including six variables, which are found to have significance for COVID-19 mortality of patients (Table 7.2). The insignificant value of Hosmer-Lemeshow's goodness of fit guides us to accept the null hypothesis, i.e., no difference between observed and model-predicted values.

Table 7.3 shows the results of binary logistic regression analyses for COVID-19 mortality as non-survivors versus survivors. LGS was found in 229 (51.5%) and 325 (18.8%) patients of the non-survivors and survivors' groups, respectively ($p < 0.001$). When age, gender, BMI, smoking status, presence of comorbidities, low grip strength, MET, and abnormal laboratory findings were taken into analyses (Table 7.3), COPD, LGS, SaO_2, MET, age, and CRP levels were found to be independent predictors for COVID-19 mortality (all $p < 0.05$).

7.4 DISCUSSIONS

HGS and MET play crucial roles in determining an individual's motor skill performance. As fundamental components of overall muscular function, these parameters significantly influence the ability to perform various activities and tasks effectively, either in activities of daily living or simply in any occupational tasks. HGS, reflecting the force generated by the muscles in the hand and forearm, is vital for executing a range of daily tasks, from lifting objects to writing and typing. A strong grip is associated with enhanced control and precision in hand movements, thereby improving overall motor performance. Similarly, MET, defined as the duration an individual can sustain their grip, plays a crucial role in the ability to perform repetitive or prolonged tasks, such as carrying groceries or using tools, without succumbing to muscle fatigue. These factors have been shown to be foundational motor skills essential for maintaining productivity and efficiency in the workforce (41).

In this retrospective study, we aimed to evaluate the HGS and muscular MET in COVID-19 patients admitted to ICU and to investigate the potential relationship between these measures and COVID-19 mortality. To the best of our knowledge, this is the first study to specifically investigate the connection between muscle health and function in predicting mortality among COVID-19 patients. Although some existing studies have touched upon the significance of muscle strength and mass in relation to hospital stays (42, 43), functional outcomes, and organ impairments in COVID-19 patients, none have explicitly examined the predictive value of muscle health and function in relation to mortality.

7.4.1 Age, BMI, Gender, and Smoking Effects

The relationship between age and COVID-19 mortality in this study followed a linear pattern, consistent with most previous studies (23, 44), showing a steep increase in mortality rates among patients aged 60 and older. Aging is known to be associated with declines in hormones that regulate muscle mass, such as dehydroepiandrosterone, testosterone, insulin-like growth factor I (IGF-I), and estrogen (45). Additionally, chronic elevation of inflammatory biomarkers like IL-6, TNF-α, and CRP in older adults ($\geq$65 years) leads to anabolic resistance, reducing muscle protein synthesis (MPS) during infectious outbreaks (46). These factors are linked to the higher ICU stay rates in older adults, who often have comorbidities such as CVD, COPD, DM, cancer, sarcopenia, cachexia, and lower quality of life, all of which increase their mortality risk (47).

Our study findings revealed a higher mortality rate among overweight and obese patients compared to those with a healthy weight, which is consistent with previous studies (48, 49). The increased mortality among overweight or obese patients may be explained by factors such as reduced chest wall compliance and respiratory muscle strength, leading to airway narrowing and obstruction (48).

Our findings also align with global COVID-19 studies, which have shown a male bias in COVID-19 mortality, with males experiencing a higher mortality rate than females (50, 51). This disparity may be attributable to a higher prevalence of risky behaviors (e.g., smoking, alcohol consumption) and a greater incidence of comorbidities (e.g., CVD, diabetes, hypertension, cancer) among males compared to females, contributing to increased COVID-19 severity and mortality (52, 53).

In agreement with the previous studies (54, 55), our study showed that mortality was 1.5 times higher among smokers than non-smokers hospitalized with COVID-19 in the ICU, regardless of variant type. A recent meta-analysis confirmed that smokers have 1.9 times the odds of COVID-19 disease progression compared to non-smokers (56). Smoking is known to impair immune function by reducing the efficacy of macrophage cytokines, thereby weakening the body's defense against infections and almost doubling the risk of developing tuberculosis (57).

7.4.2 Comorbidities and Clinical Symptoms Effects

The findings of this study indicate that COVID-19 poses a disproportionate risk to patients in all age groups, particularly those with at least one underlying comorbidity (58). The most common symptoms in our study were fever (44.6%), cough (44.2%), and dyspnea (21.9%), which align with those reported in similar studies (59, 60). Consistent with previous research, our results confirm that elderly patients (aged ≥60) with pre-existing conditions such as hypertension, DM, obesity, bronchial asthma, COPD, and CVD are more likely to experience severe outcomes and even higher mortality in the ICUs (61, 62). Specifically, hypertension (37.5%), DM (18.1%), cardiovascular disease (15.7%), and obesity (13.1%) were the most prevalent comorbidities among the deceased elderly patients in this study. Further analysis by using binary logistic regression revealed that COPD was the most significant predictor of mortality in patients infected with COVID-19. Emerging evidence suggests that increased expression of ACE2 receptors, which are associated with COPD, exacerbates the severity of SARS-CoV infection by promoting the production of inflammatory cytokines such as IL-6, IL-8, and IL-10, which are linked to ARDS (63).

Our study also identified elevated levels of CRP, ferritin, D-dimer, WBC, and neutrophils as key factors associated with increased mortality among patients with COVID-19. Conversely, deceased patients had significantly lower levels of SaO_2, thrombocytes, and hemoglobin compared to survivors. These findings are consistent with prior research by Yang et al., She et al., and Rostamzadeh et al., who reported similar associations between abnormal clinical and laboratory markers and more severe COVID-19 outcomes, particularly in older patients (21, 25, 29). However, in our binary logistic regression model, only elevated CRP and decreased SaO_2 were independently linked to mortality in ICU patients. Vetrugno et al. noted that although invasive or non-invasive ventilation may improve SaO_2 levels, it can also increase the risk of barotrauma and mortality, especially in patients with severe COVID-19 (64).

7.4.3 LGS and MET Effects

Our findings suggest that in addition to well-established clinical risk factors, LGS and reduced muscular MET can be considered independent predictors of COVID-19 mortality in patients. The study revealed that deceased patients, regardless of gender, exhibited lower grip strength and muscular MET, as well as a higher prevalence of LGS compared to survivors. In line with the present study, Kara et al. found that LGS is a significant factor in predicting the severity of COVID-19 in patients aged 21–74 years (20). Similarly, studies by Sevilla et al. and Vincenzo Galluzzo concluded that both low HGS and reduced muscular MET independently contributed to longer ICU stays and greater disease severity in elderly COVID-19 patients (65, 66). Msigwa et al. suggested that increased protein turnover in COVID-19 patients, triggered by intense proteolytic activity, can result in direct muscle inflammation caused by equine coronavirus (ECoV) particles and immune-mediated muscle injury (67). Recent studies have demonstrated that muscle mass decline, commonly observed in severe COVID-19 cases, is linked to lower levels of testosterone and IGF-I, both of which contribute to muscle loss (68, 69). Additionally, isolation/quarantine or hospitalization during the COVID-19 pandemic has been shown to accelerate the decline in isometric and concentric strength, reduce the number and size of muscle fibers, and worsen physical abilities and severity of disease in patients (47, 70).

The present study on HGS in COVID-19 patients has several limitations that warrant consideration in future research. First, the absence of detailed patient medical histories prior to ICU admission may have influenced the findings, as these data could offer valuable insights into predisposing factors affecting HGS and neuromotor control. Second, variables like dietary habits, lifestyle, physical activity levels, anthropometric data, maturity stages, COVID-19 variants, and SARS-CoV-2 vaccination were not considered in this study, potentially limiting the comprehensiveness of the results. Future studies should include these factors to minimize misinterpretation. Third, the study lacked detailed information on patients' oxygen intake process and its effects on neuromotor control and muscle strength. Future studies should consider monitoring oxygen saturation levels and standardizing oxygen intake during HGS measurements to enhance accuracy and comparability. Despite these limitations, this study highlights the clinical relevance of HGS assessment in COVID-19 patients. Future research addressing these gaps may enhance understanding of the interplay between HGS, neuromotor control, and COVID-19, enabling clinicians to develop more effective strategies for evaluating and improving muscle strength and functional outcomes in this patient population.

7.5 CONCLUSIONS

This study enhances our understanding of the prevalence and prognostic significance of LGS and MET in COVID-19 patients admitted to the ICU. HGS and MET are significant factors in assessing young and adult motor skills. Alongside established clinical and laboratory risk factors such as age, COPD, CRP, and SaO_2, HGS and MET – key indicators of neuromuscular health – emerge as quick and low-cost prognostic tools for predicting COVID-19 mortality. Our findings suggest that reduced grip strength and MET could serve as important risk factors, assisting clinicians in early diagnosis, risk stratification, and follow-up of COVID-19 patients. This approach could enhance the effectiveness of clinical interventions during the pandemic while optimizing the use of financial and human resources.

ACKNOWLEDGMENT

This work was supported by the Shahid Beheshti University of Medical Sciences, Tehran, Iran (under Grant No. 43005817). The authors would like to express special thanks to all the physicians as well as patients and their families who chose to be involved in this research.

REFERENCES

1. Saremi M, Rostamzadeh S. Hand dimensions and grip strength: a comparison of manual and non-manual workers. In: *Congress of the International Ergonomics Association*. Springer, Florence, Italy. 2018. pp. 520–529.
2. Soysal P, Hurst C, Demurtas J, Firth J, Howden R, Yang L, et al. Handgrip strength and health outcomes: umbrella review of systematic reviews with meta-analyses of observational studies. *J Sport Heal Sci*. 2021;10(3):290–295.
3. Soyuer F, Cankurtaran F, Menevşe Ö, Ertürk Zararsız G. Examination of the correlation between hand grip strength and muscle mass, balance, mobility, and daily life activities in elderly individuals living in nursing homes. *Work*. 2023 Jan 1;74(4):1371–1378.
4. Malhotra R, Tareque MI, Tan NC, Ma S. Association of baseline hand grip strength and annual change in hand grip strength with mortality among older people. *Arch Gerontol Geriatr*. 2020;86(1):103961.
5. Saremi M, Rostamzadeh S, Nasr Esfahani M. Hand functionality in dentists: the effect of anthropometric dimensions and specialty. *Int J Occup Saf Ergon*. 2021; 28(3):1473–1481.
6. Rostamzadeh S, Abouhossein A, Alam K, Vosoughi S, Sattari SS. Exploratory analysis using machine learning algorithms to predict pinch strength by anthropometric and socio-demographic features. *Int J Occup Saf Ergon*. 2024; 30(2):518–531.
7. Rostamzadeh S, Saremi M, Tabatabaei S. Normative hand grip strength and prediction models for Iranian office employees. *Work*. 2019;62(2):233–241.
8. Görner K, Reineke A. The influence of endurance and strength training on body composition and physical fitness in female students. *J Phys Educ Sport*. 2020;20(1):2013–2020.
9. Jakobsen LH, Rask IK, Kondrup J. Validation of handgrip strength and endurance as a measure of physical function and quality of life in healthy subjects and patients. *Nutrition*. 2010;26(5):542–550.
10. Eksioglu M. Endurance time of grip-force as a function of grip-span, posture and anthropometric variables. *Int J Ind Ergon*. 2011;41(5):401–409.
11. Rostamzadeh S, Abouhossein A, Allafasghari A, Allafasghari A. Importance of handgrip strength and endurance time for predicting COVID-19 mortality in older adult patients: K-Nearest Neighbors (K-NN) algorithm. *AHFE. Nice, France*. 2024;133(1):116–127.
12. Rostamzadeh S, Saremi M, Taheri F. Maximum handgrip strength as a function of type of work and hand-forearm dimensions. *Work*. 2020; 65(3):679–687.
13. Neri SGR, Lima RM, Ribeiro HS, Vainshelboim B. Poor handgrip strength determined clinically is associated with falls in older women. *J Frailty, Sarcopenia Falls*. 2021;6(2):43.
14. Jimenez KA, Kwon J-W, Yoon J, Lee H-M, Moon S-H, Suk K-S, et al. Handgrip strength correlated with falling risk in patients with degenerative cervical myelopathy. *J Clin Med*. 2021;10(9):1980–1992.
15. Spiering BA, Mujika I, Sharp MA, Foulis SA. Maintaining physical performance: the minimal dose of exercise needed to preserve endurance and strength over time. *J Strength Cond Res*. 2021;35(5):1449–1458.
16. Aisbett B, Nichols D. Fighting fatigue whilst fighting bushfire: an overview of factors contributing to firefighter fatigue during bushfire suppression. *Aust J Emerg Manag*. 2007;22(3):31–39.
17. de la Motte SJ, Gribbin TC, Lisman P, Murphy K, Deuster PA. Systematic review of the association between physical fitness and musculoskeletal injury risk: Part 2 – Muscular endurance and muscular strength. *J Strength Cond Res*. 2017;31(11):3218–3234.

18. Ali AM, Kunugi H. Physical frailty/sarcopenia as a key predisposing factor to coronavirus disease 2019 (COVID-19) and its complications in older adults. *BioMed.* 2021;1(1):11–40.
19. Grigioni S, Lvovschi V-E, Tamion F, Joly L-M, Coëffier M, Van Elslande H, et al. Low thoracic skeletal muscle index is associated with negative outcomes in 244 patients with respiratory COVID-19. *Clin Nutr.* 2023;42(2):102–107.
20. Kara Ö, Kara M, Akın ME, Özçakar L. Grip strength as a predictor of disease severity in hospitalized COVID-19 patients. *Hear Lung.* 2021;50(6):743–747.
21. Rostamzadeh S, Allafasghari A, Allafasghari A, Abouhossein A. Handgrip strength as a prognostic factor for COVID-19 mortality among older adult patients admitted to the intensive care unit (ICU): a comparison Alpha (B. 1.1. 7) and Delta (B. 1.617. 2) *variants. Sci Rep.* 2024;14(1):19927.
22. Pijls BG, Jolani S, Atherley A, Derckx RT, Dijkstra JIR, Franssen GHL, et al. Demographic risk factors for COVID-19 infection, severity, ICU admission and death: a meta-analysis of 59 studies. *BMJ Open.* 2021;11(1):e044640.
23. Guan W, Ni Z, Hu Y, Liang W, Ou C, He J, et al. Clinical characteristics of coronavirus disease 2019 in China. *N Engl J Med.* 2020;382(18):1708–1720.
24. Puthucheary ZA, Rawal J, McPhail M, Connolly B, Ratnayake G, Chan P, et al. Acute skeletal muscle wasting in critical illness. *Jama.* 2013;310(15):1591–1600.
25. Yang X, Yu Y, Xu J, Shu H, Liu H, Wu Y, et al. Clinical course and outcomes of critically ill patients with SARS-CoV-2 pneumonia in Wuhan, China: a single-centered, retrospective, observational study. *Lancet Respir Med.* 2020;8(5):475–481.
26. Tang Y, Liu J, Zhang D, Xu Z, Ji J, Wen C. Cytokine storm in COVID-19: the current evidence and treatment strategies. *Front Immunol.* 2020;11(4):1708–1722.
27. Files DC, Ilaiwy A, Parry TL, Gibbs KW, Liu C, Bain JR, et al. Lung injury-induced skeletal muscle wasting in aged mice is linked to alterations in long chain fatty acid metabolism. *Metabolomics.* 2016;12(4):1–16.
28. Zhou J, Liu B, Liang C, Li Y, Song Y-H. Cytokine signaling in skeletal muscle wasting. *Trends Endocrinol Metab.* 2016;27(5):335–347.
29. She Q, Chen B, Liu W, Li M, Zhao W, Wu J. Frailty pathogenesis, assessment, and management in older adults with COVID-19. *Front Med.* 2021;8(6):694367.
30. Macedo A, Goncalves N, Febra C. COVID-19 fatality rates in hospitalized patients: systematic review and meta-analysis. *Ann Epidemiol.* 2021;57(4):14–21.
31. Izumi T, Das S, Abe M, Shaw R. Managing compound hazards: impact of COVID-19 and cases of adaptive governance during the 2020 Kumamoto flood in Japan. *Int J Environ Res Public Health.* 2022;19(3):1188.
32. World Health Organization. International Association for the Study of Obesity, International Obesity Task Force. *The Asia-Pacific perspective: redefining obesity and its treatment.* Sydney: Health Communications Australia; 2000; pp. 54–134.
33. Valenza F, Guglielmi M, Maffioletti M, Tedesco C, Maccagni P, Fossali T, et al. Prone position delays the progression of ventilator-induced lung injury in rats: does lung strain distribution play a role? *Crit Care Med.* 2005;33(2):361–367.
34. Hansell DM, Bankier AA, MacMahon H, McLoud TC, Muller NL, Remy J. Fleischner Society: glossary of terms for thoracic imaging. *Radiology.* 2008;246(3):697–722.
35. Pan F, Ye T, Sun P, Gui S, Liang B, Li L, et al. Time course of lung changes at chest CT during recovery from coronavirus disease 2019 (COVID-19). *Radiology.* 2020;295(3):715–721.
36. Richards L, Palmiter-Thomas P. Grip strength measurement: a critical review of tools, methods, and clinical utility. *Crit Rev Phys Rehabil Med.* 1996;8(1–2):87–109.

37. Dodds RM, Syddall HE, Cooper R, Benzeval M, Deary IJ, Dennison EM, et al. Grip strength across the life course: normative data from twelve British studies. *PLoS One*. 2014;9(12):e113637.
38. Beaudart C, Rolland Y, Cruz-Jentoft AJ, Bauer JM, Sieber C, Cooper C, et al. Assessment of muscle function and physical performance in daily clinical practice: a position paper endorsed by the European Society for Clinical and Economic Aspects of Osteoporosis, Osteoarthritis and Musculoskeletal Diseases (ESCEO). *Calcif Tissue Int*. 2019;105(5):1–14.
39. Grubbs FE. Procedures for detecting outlying observations in samples. *Technometrics*. 1969;11(1):1–21.
40. Singh SG, Kumar SV. Dealing with multicollinearity problem in analysis of side friction characteristics under urban heterogeneous traffic conditions. *Arab J Sci Eng*. 2021;46(4):10739–10755.
41. Singh GK, Srivastava S. Grip strength of occupational workers in relation to carpal tunnel syndrome and individual factors. *Int J Occup Saf Ergon*. 2020;26(2): 296–302.
42. Bashirian S, Jenabi E, Khazaei S, Barati M, Karimi-Shahanjarini A, Zareian S, et al. Factors associated with preventive behaviours of COVID-19 among hospital staff in Iran in 2020: an application of the Protection Motivation Theory. *J Hosp Infect*. 2020;105(3):430–433.
43. Roth GA, Emmons-Bell S, Alger HM, Bradley SM, Das SR, De Lemos JA, et al. Trends in patient characteristics and COVID-19 in-hospital mortality in the United States during the COVID-19 pandemic. *JAMA Netw Open*. 2021;4(5):e218828–e218828.
44. Huang C, Wang Y, Li X, Ren L, Zhao J, Hu Y, et al. Clinical features of patients infected with 2019 novel coronavirus in Wuhan, China. *Lancet*. 2020;395(10223):497–506.
45. Vitale G, Cesari M, Mari D. Aging of the endocrine system and its potential impact on sarcopenia. *Eur J Intern Med*. 2016;35(6):10–15.
46. Singh T, Newman AB. Inflammatory markers in population studies of aging. *Ageing Res Rev*. 2011;10(3):319–329.
47. Garg S, Kim L, Whitaker M, O'Halloran A, Cummings C, Holstein R, et al. Hospitalization rates and characteristics of patients hospitalized with laboratory-confirmed coronavirus disease 2019 – COVID-NET, 14 States, March 1–30, 2020. *Morb Mortal Wkly Rep*. 2020;69(15):458–464.
48. Soeroto AY, Soetedjo NN, Purwiga A, Santoso P, Kulsum ID, Suryadinata H, et al. Effect of increased BMI and obesity on the outcome of COVID-19 adult patients: a systematic review and meta-analysis. *Diabetes Metab Syndr Clin Res Rev*. 2020;14(6):1897–1904.
49. Al-Salameh A, Lanoix J-P, Bennis Y, Andrejak C, Brochot E, Deschasse G, et al. The association between body mass index class and coronavirus disease 2019 outcomes. *Int J Obes*. 2021;45(3):700–705.
50. Gebhard C, Regitz-Zagrosek V, Neuhauser HK, Morgan R, Klein SL. Impact of sex and gender on COVID-19 outcomes in Europe. *Biol Sex Differ*. 2020;11(4):1–13.
51. Brady E, Nielsen MW, Andersen JP, Oertelt-Prigione S. Lack of consideration of sex and gender in COVID-19 clinical studies. *Nat Commun*. 2021;12(1):4015–4021.
52. Abate BB, Kassie AM, Kassaw MW, Aragie TG, Masresha SA. Sex difference in coronavirus disease (COVID-19): a systematic review and meta-analysis. *BMJ Open*. 2020;10(10):e040129.
53. Singh S, Chowdhry M, Chatterjee A, Khan A. Gender-based disparities in COVID-19 patient outcomes: a propensity-matched analysis. *MedRxiv*. 2020; 2004–2020.
54. Karanasos A, Aznaouridis K, Latsios G, Synetos A, Plitaria S, Tousoulis D, et al. Impact of smoking status on disease severity and mortality of hospitalized patients

with COVID-19 infection: a systematic review and meta-analysis. *Nicotine Tob Res.* 2020;22(9):1657–1659.

55. Alqahtani JS, Oyelade T, Aldhahir AM, Alghamdi SM, Almehmadi M, Alqahtani AS, et al. Prevalence, severity and mortality associated with COPD and smoking in patients with COVID-19: a rapid systematic review and meta-analysis. *PLoS One.* 2020;15(5):e0233147.
56. Patanavanich R, Glantz SA. Smoking is associated with COVID-19 progression: a meta-analysis. *Nicotine Tob Res.* 2020;22(9):1653–1656.
57. López-Hernández Y, Rivas-Santiago CE, López JA, Mendoza-Almanza G, Hernandez-Pando R. Tuberculosis and cigarette smoke exposure: an update of in vitro and in vivo studies. *Exp Lung Res.* 2018;44(2):113–126.
58. Flaherty GT, Hession P, Liew CH, Lim BCW, Leong TK, Lim V, et al. COVID-19 in adult patients with pre-existing chronic cardiac, respiratory and metabolic disease: a critical literature review with clinical recommendations. *Trop Dis Travel Med vaccines.* 2020;6(8):1–13.
59. Kabeerdoss J, Pilania RK, Karkhele R, Kumar TS, Danda D, Singh S. Severe COVID-19, multisystem inflammatory syndrome in children, and Kawasaki disease: immunological mechanisms, clinical manifestations and management. *Rheumatol Int.* 2021;41(1):19–32.
60. Gallo Marin B, Aghagoli G, Lavine K, Yang L, Siff EJ, Chiang SS, et al. Predictors of COVID-19 severity: a literature review. *Rev Med Virol.* 2021;31(1):1–10.
61. Onder G, Rezza G, Brusaferro S. Case-fatality rate and characteristics of patients dying in relation to COVID-19 in Italy. *Jama.* 2020;323(18):1775–1776.
62. Woolf SH, Chapman DA, Lee JH. COVID-19 as the leading cause of death in the United States. *Jama.* 2021;325(2):123–124.
63. He L, Ding Y, Zhang Q, Che X, He Y, Shen H, et al. Expression of elevated levels of pro-inflammatory cytokines in SARS-CoV-infected ACE2+ cells in SARS patients: relation to the acute lung injury and pathogenesis of SARS. *J Pathol A J Pathol Soc Gt Britain Irel.* 2006;210(3):288–297.
64. Vetrugno L, Castaldo N, Fantin A, Deana C, Cortegiani A, Longhini F, et al. Ventilatory associated barotrauma in COVID-19 patients: A multicenter observational case control study (COVI-MIX-study). *Pulmonology.* 2023;29(6):457–468.
65. Galluzzo V, Zazzara MB, Ciciarello F, Savera G, Pais C, Calvani R, et al. Fatigue in Covid-19 survivors: the potential impact of a nutritional supplement on muscle strength and function. *Clin Nutr ESPEN.* 2022;51(10):215–221.
66. Sevilla GGP de, Sánchez-Pinto B. Associations between muscle strength, dyspnea and quality of life in post-COVID-19 patients. *Rev Assoc Med Bras.* 2022;68(12):1753–1758.
67. Msigwa SS, Wang Y, Li Y, Cheng X. The neurological insights of the emerging coronaviruses. *J Clin Neurosci.* 2020;78(8):1–7.
68. Maggio M, Basaria S, Ceda GP, Ble A, Ling SM, Bandinelli S, et al. The relationship between testosterone and molecular markers of inflammation in older men. *J Endocrinol Invest.* 2005;28(11):116–119.
69. Wang X, Sun H, Ma B, Gao J, Yin J, Qu S. Insulin-like growth factor 1 related to chronic low-grade inflammation in patients with obesity and early change of its levels after laparoscopic sleeve gastrectomy. *Obes Surg.* 2020;30(9):3326–3332.
70. Filgueira TO, Castoldi A, Santos LER, de Amorim GJ, de Sousa Fernandes MS, Anastácio W de L do N, et al. The relevance of a physical active lifestyle and physical fitness on immune defense: mitigating disease burden, with focus on COVID-19 consequences. *Front Immunol.* 2021;12:587146.

8 A Policy Approach to Safety Incentives Implementation for Construction Safety Practices in Developing Countries

Kunle E. Ogundipe, Babatunde F. Ogunbayo, Clinton Aigbavboa and Olatoyese Z. Oni

8.1 INTRODUCTION

Workers' health and safety (H&S) practices have long been a global concern, as the construction industry globally remains one of the riskiest sectors, particularly in developing countries (Ogundipe, Ogunde, et al., 2018; Boadu et al., 2021). Othman et al. (2019) and Adekunle et al. (2023) argued that the industry is known for safety concerns globally, resulting in thousands of occupation-related deaths and injuries experienced annually across various sectors. Kalatpour and Khavajiv (2016) and Abukhashabah et al. (2020) noted that in developing countries, the construction workplace accident records are higher than international standards due to manual operation and exposure to hazardous working conditions. The Ghanaian construction industry recorded the highest number of incidents of occupational accidents, more than other industries, while 56% out of 902 recorded incidents in the year 2000 resulted in death (Osei-Asibey et al., 2021). The Ghana Statistical Service (GSS, 2016) and Osei-Asibey et al. (2021) indicated that the percentage of the construction industry in the national indicators of occupational injury is 4.7%, with a record of 65 death rates, 151% higher than the national indicator in 2015. Moda et al. (2023) asserted that about 5–7% of the workforce in the Nigerian construction industry is lost annually to construction accidents, while many people leave the industry owing to a lack of incentives to motivate the workforce. Mahmoud, Ahmad, and Yatim (2020) added that the possibility of about 2% of accidents and 5% of injury levels are recorded among construction workers in Nigeria. As a result, safety has emerged as a new metric for assessing the performance of construction projects in addition to the time, cost, and quality triangle, and it must be improved (Ngacho & Das, 2014; Ogundipe, Olanirah, et al., 2018; Oni et al., 2023).

 DOI: 10.1201/9781003672036-8

Consequently, Ahmed and Faheem (2021) and Oni et al. (2024) maintained that the need to improve workplace safety behaviours and practices has become dominant in the construction sector in developing countries. This often supports the International Labour Organisation Convention ([ILO, 2006]) advocating for standard policies to protect employees from occupational fatalities and injuries, which adversely influence national productivity and performance and social and economic development. Further, goals 3 and 8 of the United Nations Sustainable Development Goals support organisations to promote value for decent work and ensure workplace safety and workers' well-being across various occupational sectors (Moda et al., 2023). The Africa Agenda 2063 emphasises the innovative framework to attain sustainable and inclusive development and economic growth in the continent with a priority on workplace safety as part of goal 1 for the continent's 50 years of development trajectory (Ndizera & Muzee, 2018; Nwozor et al., 2021; Moda et al., 2023). Nonetheless, despite substantial efforts towards improving workplace safety and workers' well-being through institutional safety policies and academic research, safety-related challenges to the workplace persist, particularly among workers employed in hazardous industries, including construction and mining (Wu et al., 2018; Afolabi et al., 2021; Zailani et al., 2021). The phenomenon is considered worst in developing industries due to ineffective H&S policies, lack of safety incentive policies, poor H&S programmes, non-compliance with safety laws and regulations, and weak enforcement frameworks (Abubakar et al., 2022; Osei-Asibey et al., 2021).

Given the high construction-related accident and fatality rate, implementing safety incentive schemes could reduce these incidents, increase productivity, and lower costs associated with accidents and injuries. Ogwueleka (2015) and Liu et al. (2022) maintained that safety incentives are integral to the construction organisational policy to improve H&S culture, safety practices, and performance-related goals. Construction organisations are now exploring using various incentive schemes to align workers with performance goals (Ogwueleka, 2015; Ogundipe et al., 2024a). Safety incentives offer proactive rather than reactive approaches to enhance organisational H&S culture, safety practices, and performance to address safety challenges (Adekunle et al., 2023; Ogundipe et al., 2024). According to Liu et al. (2022) and Ogundipe et al. (2024), safety incentives (SI) are schemes to improve workplace H&S practices, share project risk between clients and contractors, and align workers' incentive drives with organisational performance-related goals.

This study defines safety incentive policy as a mix of policy strategies and instruments delineating basic guidelines, standard procedures, and templates for designing and implementing safety incentives within the construction industry (Ogundipe et al., 2024b). Kankaanpää (2010) posited that safety incentives are designed as a policy tool employers or construction firms use to promote occupational safety and health (OSH) among stakeholders. This provides a policy approach to safety incentives implementation to improve H&S planning, workplace safety behaviour, practices, and performance that often affect construction activities in developing countries (Ahmed & Faheem, 2021; Ogundipe et al., 2024b). Hence, Awwad et al. (2016) noted that the non-implementation of H&S regulatory policies and lack of understanding of drivers of safety incentives policy negatively affect the construction industry's implementation of safety incentives to support H&S practices. Thus, addressing the gap in knowledge of drivers of policy approach to safety incentives design is imperative, which could significantly influence the implementation of safety incentives in the construction sector in developing economies, particularly in Nigeria. Given this, the study aims to identify drivers of policy approach to safety incentives implementation through construction professionals' perspectives to enrich its application among construction organisations in developing economies. A review of extant literature was conducted to understand the drivers of the policy approach to safety incentives implementation in the construction industry and established drivers that significantly influence the Nigerian construction sector to achieve this aim.

8.2 SAFETY INCENTIVES POLICY IN THE CONSTRUCTION INDUSTRY

Safety incentives provide various benefits that include lower accident and injury rates, lower material and equipment damage rates, lower absenteeism rates, higher motivation, enhanced safety performance, higher employee performance, fostering positive attitudes in the workplace, and increased job satisfaction (Fernåndez-Muiz et al., 2009; Moosa & Oriet, 2022). According to Del-Puerto and Elliott (2012) and Ogundipe et al. (2024), the policy approach to safety incentives implementation drives construction firms and stakeholders to promote safe workplace behaviour and culture, facilitate equipment and resources that aid safety tasks in the workplace, and provide a clear understanding of approaches to incentives selection methods. Ward et al. (2016) admitted that safety incentive policies could be performance-based, individual and team, profit sharing and co-ownership, and direct and indirect incentive policies. A performance-based SI policy is based on task outcome indicators directly influencing construction organisations or project goals (Ward et al., 2016). This type of incentive policy allows construction firms to design bonuses for individual employees or groups of employees for completing projects based on the performance indicators (safety, quality, time, or within budget) established before the commencement of the task/project. A performance-based safety incentive policy may be included in the contractual arrangement so that contracting firms can meet project requirements proactively regarding time, cost, and quality (Ogwueleka, 2015).

Ogwueleka (2015) admitted that a performance-based safety incentives policy often includes bonuses and penalties payable for meeting the performance indicators. Ogwueleka (2015) and Ogunbayo et al. (2024) contended that the performance-based safety incentive policy is based on performance parameters, incentive structure, evaluation period, and risk/benefit sharing formulae. Lee and Kim (2024) applied image-based activity classification to examine how safety incentives affect worker safety behavioural control. The study reveals that it is more effective to reward good compliance when there is no feedback on compliance status and to penalise non-conformity when there are three feedbacks during the jobs (Lee & Kim, 2024). The performance-based approach provides an accurate assessment tool for safety incentives because it focuses on improving the performance of safe behaviours to promote safety compliance among construction workers. Similarly, a team or individual safety incentive is an employee-focused safety incentive policy. This incentive policy provides an understanding of the use of safety incentives, mainly when there is interdependent work and the need for employees to work together (Martocchio, 2015). The policy approach often promotes a collaborative spirit and safety behaviour among employees, cultivating loyalty to an organisation and improving productivity and quality delivery (Martocchio, 2015). Bauer and Erdogan (2012) stated that individual safety incentive policies could be one-time payments that follow employees' specific accomplishments.

The profit-sharing and co-ownership incentives policy approach to incentive implementation involves a percentage sharing of the organisations' profits with their workers. The profit-sharing safety incentive policy approach, as described by De-Silva (2004) and Bryson et al. (2011), provides payment schemes that qualify workers to share a cash bonus or a scheduled payment kept in a trust fund or share options. This policy allows organisations to implement rewards based on an agreed sharing ratio among employees and employers linked to specific performance improvements and productivity gains (De Silva, 2004; Bryson et al., 2011). The co-ownership incentive policy approach allows workers to own stock or shares within their organisations in exchange for performance bonuses (Enters et al., 2003). A direct incentive policy is designed to influence organisational resources and return to investment directly and instantly (Enters et al., 2003). Enters et al. (2003) noted that the direct incentive policy provision includes the following: materials subsidy, grant enabling H&S practices, government tax relief, access to scarce resources, subsidised loans, and price guarantees and cost-sharing arrangements. Enters et al. (2003) described indirect incentives as elements of an incentive policy that do not have a direct effect by changing or setting the overall policy conditions within and outside the organisations. According to Enters et al. (2003) and the Food and Agriculture Organisation of the United Nations (FAO, 1999), an indirect incentive policy could be variable incentives associated with economic factors affecting construction organisations' activities (H&S practices) and enabling incentive factors prompting service providers to mediate variable incentives. Hence, the indirect incentive policy is designed to offer resource security, national security, research and development, political and macroeconomic stability, credit facilities, and service provider support, enabling organisations to thrive (Food and Agriculture Organisation of the United Nations [FAO], 1999). Similarly, variable incentives are other economic factors, such as exchange rate, interest rate, general taxes, and fiscal and monetary measures, that indirectly influence construction industry operations (Enters et al., 2003).

8.3 SAFETY INCENTIVE IMPLEMENTATION MECHANISMS FOR CONSTRUCTION SAFETY PRACTICES

Attaining safety compliance on construction sites has long been a top priority, and researchers often identify various improvement strategies through reactive and proactive approaches to safety management (Ogundipe, Olanirah, et al., 2018; Workman, 2018; Adekunle et al., 2023). The reactive approach to safety aims to respond to incidents as they occur and take measures to prevent similar incidents in the future.

The proactive approach involves early detection and prevention strategies for potential accidents and provides warning of fatal events (Adekunle et al., 2023). This section focuses on a proactive approach to construction safety practices and management using a safety incentives policy approach to enhance safety practices and align workers with organisational performance-related goals. To this end, safety incentives are applied in the construction sector to attain project objectives by setting measurable performance-related goals using elements that impact the project schedule, quality, and safety practices (Alfandi & Alkahsawneh, 2014; Ogwueleka, 2015; Ogundipe et al., 2024). Ahmed and Faheem (2021) and Ogundipe et al. (2025) admitted that reward and penalty schemes are used in construction contracts to align firms with performance-related goals, share risk between clients and contractors, and enhance employees with workplace safety practices. Previous studies have established techniques to improve the implementation of safety incentives to benefit construction stakeholders in mutual contractual arrangements. Imhof and Kräkel (2014) assessed the principles of moral hazard problems in contracts and that implementing bonus contracts provides individual stakeholders with incentives for meeting a particular performance threshold.

Wang et al. (2023), on the other hand, noted that effective safety incentive implementation is supported by the technology integration in incentive mechanisms and the design process of optimal incentive contracts using various performance indicators. Nadari et al. (2023) developed an automation process: Fungible Tokens and Non-Fungible Tokens for automated incentive mechanisms in incentive contracts, offering a decentralised, transparent, traceable, and immutable incentive mechanism to improve stakeholder trust. Bao et al. (2024) noted that integrating web- and blockchain-based incentives improves the implementation of safety incentives using a native token incentive mechanism. The study established that linking workplace safety requirements with web- and blockchain-based incentives could motivate workers to regularly participate in safety training and enhance safety awareness and knowledge, cultivating a new safety culture (Bao et al., 2024). Ji, Liu, and Zhang (2021) and Zhu et al. (2022) explore evolutionary game analysis and tournament mechanism dimensions in safety incentives implementation in the construction industry. However, Ji, Liu, and Zhang (2021) noted that the challenges of individual workers' fairness preference, risk preference, and ability level in safety incentive implementation could be overcome by integrating the competition tournament rewards mechanism. Moreover, Zhu et al. (2022) explore an evolutionary game model in developing incentive and punishment mechanisms to improve the effectiveness of workplace safety practices between contractors and construction workers. The study of Zhu et al. (2022) established the impact of the evolutionary game analysis model as a mechanism for promoting mutual trust, benefits allocation, and setting incentives, punishment and reward indicators mutually beneficial to contractors and workers. However, various studies have established mechanisms for efficiently implementing safety incentives within the construction industry (Ji et al., 2021; Zhu et al., 2022; Bao et al., 2024; Wang et al., 2023). Unfortunately, the knowledge gap of drivers of policy approach to safety incentives design significantly affects the choice of construction firms' implementation of safety incentives in developing countries, including Nigeria.

8.4 POLICY DRIVERS FOR SAFETY INCENTIVES IMPLEMENTATION IN THE CONSTRUCTION INDUSTRY

Implementing a safety incentives policy in the construction sector demonstrates its impact on effectively improving workplace safety practices and meeting workers' incentive drives (Boadu et al., 2021; Ogundipe et al., 2024d). This is due to safety incentive policy drivers that influence construction firms to implement incentive schemes to improve workplace safety practices and organisational performance-related goals. Dambisya and Dambisya (2007) attributed that non-financial and financial schemes are often implemented in Eastern and Southern African countries to incentivise workers with the aim to enhance workplace conditions, employee training, capacity development, social support, salary enhancements and allowances, and access to healthcare. Likewise, Ghasemi et al. (2015) noted that the driver of incentive policy in Iran lies in its ability to stir behavioural changes in employees, provide guidelines for safety incentive design, and provide KPIs assessment procedures for frontline employees in the construction industry. Kigule (2013) admitted that the pay and incentive policy implemented by the Tanzania government supports organisations to use standard and transparent incentive schemes, design appropriate rewards and penalties for acceptable performance standards, and plan and control employees' wages. Choi et al. (2012) attributed the importance of the pay-for-safety incentive schemes designed for public works construction contracts as drivers for the effective implementation of safety incentives to enhance safety performance in Hong Kong.

Similarly, the research by Wilde (2011) and Jiang et al. (2019) indicated that policies supporting incentive goal settings and performance indicators influence the implementation of safety incentives by construction firms in developing countries. According to Abubakari et al. (2018), drivers of implementing incentive policy are essential for organisations or countries to plan, regulate, control, and guide their growth and development. Ward et al. (2016) and Ogundipe et al. (2024b) admitted that incentive policy implementation must balance many interrelated motivating factors essential for maximising personnel and resource management and establishing incentive selection approaches. Hassan and Jha (2013) identify six essential drivers for implementing safety incentive policy as follows: approaches to incentive selection method, understanding H&S workplace conditions of work, employee training, the role of the safety committee and subcontractors, specialised works and equipment safety, standalone H&S incentive policy, and appropriate incentive and penalty design. Howlett (2019) added that drivers of incentive policymaking require creating and implementing mixes of policy instruments for incentive budget guidelines and accelerating economic growth expected to achieve policy objectives. Del-Puerto and Elliott (2012) noted that the incentives policy for construction firms intends to drive long-term H&S national development, establish incentive selection approaches, and improve work conditions among workers.

Hence, Table 8.1 summarises the drivers that construction firms and stakeholders can easily understand when designing and implementing safety incentives.

TABLE 8.1
Policy Drivers for Safety Incentives Implementation

Policy Drivers for Safety Incentives	Authors
Incentive budget planning guidelines	Howlett (2019)
Organisational resource management	Ward et al. (2016), Ogundipe et al. (2024b)
Clear incentive goal settings	Wilde (2011), Jiang et al. (2019)
Organisational workforce management	Ward et al. (2016), Ogundipe et al. (2024b)
Enhance safety performance	Choi et al. (2012), Abubakari et al. (2018)
Provide interrelated motivating factors	Ward et al. (2016), Ogundipe et al. (2024b)
Long-term H&S development plan	Del-Puerto and Elliott (2012)
Establishing safety incentive determinants	Zhu et al. (2022)
Guideline for safety incentives design	Kigule (2013), Ghasemi et al. (2015)
Fostering workplace-positive attitudes	Fernández-Muiz et al. (2009), Moosa and Oriet (2022)
Promote workplace safety requirements	Fernández-Muiz et al. (2009), Moosa and Oriet (2022)
Proactive approach to safety management	Adekunle et al. (2023), Ogundipe et al. (2024b)
Support workers' behavioural change	Ghasemi et al. (2015)
Economy growth acceleration	Abubakari et al. (2018)
Establish incentive selection approaches	Hassan and Jha (2013), Ward et al. (2016)
Technology application in incentive mechanisms	Wang et al. (2023), Nadari et al. (2023)
Establish incentive implementation plans	Del-Puerto and Elliott (2012)
Safety incentive KPIs assessment procedures	Zhu et al. (2022)

8.5 RESEARCH METHODOLOGY

The study explored drivers of policy approach to safety incentives through a positivist approach to objectively enrich its implementation among construction organisations in developing economies. This approach has been previously adopted for several built environment studies (Adekunle et al., 2022; Owusu-Manu et al., 2023). The study adopted a quantitative approach using a questionnaire to improve the generalisability of study findings. This study was conducted in Lagos, Nigeria, using construction professionals. The study area selection is a commercial hub for various high-profile construction activities and the headquarters of most construction professional bodies in Nigeria (Ogunde, 2017). The population comprises construction professionals in the Nigerian construction-built environment. Respondents for this study were selected based on their involvement and experience with H&S practices within construction project delivery. The sample frame comprises 958 registered members of the Nigeria Institute of Architects (NIA), 610 registered members of the Nigerian Institute of Building (NIOB), 2670 registered members of the Nigeria Society of Engineers (NSE), 870 registered members of the Nigeria Institution of Quantity Surveyors (NIQS), and 300 registered members of the Nigerian Institution of Builders in Facilities Management (NIBIFM), in Lagos Chapter, Nigeria, respectively.

In determining the sample, the Yamane (1967) equation, as cited by Agrasuta and Nelson (2013), was adopted to establish the sample size representing the total population of construction professionals in the study area. Yamane (1967) noted that the level of precision is set at a 95% confidence level, which was adopted for calculating sample size in this study. Yamane's (1967) equation established 372.45 respondents as the sample size required, representing 6.88% of the sample frame of 5408. Hence, a quantitative survey questionnaire was developed using a five-point Likert scale of 1 = Strongly Disagree; 2 = Disagree; 3 = Neutral; 4 = Agree; 5 = Strongly Agree. The survey questionnaire was developed using Google Forms and emailed to the respondents through their respective state chapter offices of the NIOB, NIA, NIQS, and NSE Lagos Chapter, Nigeria, and 379 respondents were used for data analysis in this study. The study analysed data from a questionnaire survey using statistical tools such as standard deviation (SD), mean item scores, and Kruskal–Wallis tests. These tools were instrumental in exploring the perspectives of construction professionals from the survey data to gain an understanding of the drivers of policy approaches to safety incentives in the construction sector. Likewise, Pallant (2016) noted that the reliability of the data collection instrument can be determined using Cronbach's alpha to measure the scale interrelatedness of variables. A 0.889 value obtained from Cronbach's alpha for data reliability is more than the 0.6 minimum value recommended and indicates a high degree of consistency (Pallant, 2016). Thus, the rigour of the methodology and validation of data reliability ensure the study findings significantly contribute to safety incentives implementation within the construction sector in developing countries, particularly in the Nigerian context.

8.6 RESULTS

This section focuses on the respondents' background information. The survey was conducted using construction professionals' perspectives in Nigeria. The survey revealed a diverse range of designations: about 23.3% (n = 88) were architects, 32.45% (n = 123) were builders, 24.27% (n = 92) were engineers, and 20.05% (n = 76) were quantity surveyors. In addition, the results detail the respondents' professional affiliations within the Nigerian construction industry. About 23.3% (n = 88) of the respondents were members of the Nigerian Institutes of Architects (NIA), 32.45% (n = 123) were members of the Nigerian Institute of Building (NIOB), 24.27% (n = 92) were members of the NSE, and 20.05% (n = 76) were members of the NIQS. In terms of educational background, about 29.6% (n = 112) of the respondents had bachelor's degrees; 26.4% (n = 100), postgraduate diplomas; 22.2% (n = 84), master's degrees; 18.7% (n = 71), higher national diplomas; 2.2% (n = 9), doctoral degrees; and 1% (n = 3), national diplomas. To this end, the participants demonstrated vast industry experience: about 42.2% (n = 160) had between 11 and 15 years of working experience; 34.8% (n = 132) had 6–10 years of working experience; 14.5% (n = 55) had 16–20 years of working experience; 5.5% (n = 21) had 1–5 years of working experience; and 2.9% (n = 11) had more than 21 years of working experience. This distribution of the respondents' background information showcases a well-experienced and educated sample, providing a robust insight into completing the survey questions. The findings underscore that the construction professionals in this study were adequately represented and possessed the required educational background and professional competence to respond to the survey questions (Akuete et al., 2022).

The study employed the descriptive analysis of the 18 identified policy drivers for safety incentives implementation in the construction industry, as shown in Table 8.2 by ranking mean score (MS) and SD. As guided by Opawole and Jagboro's (2016) recommendations, MS values ≥ 3.50 on the five-point Likert scale suggest a high significance value of variables of policy drivers for safety incentives implementation. Hence, 13 of the 18 identified variables have mean values ≥ 3.50. Likewise, when variables have the same MS value, the variable with the lowest SD was assigned the highest priority ranking (Field, 2009; Oke et al., 2024).

The results show that enhanced safety performance is ranked first as the most significant driver (MS 4.04, SD 1.054). Organisational workforce management ranked second (MS 4.02, SD 0.962); interrelated motivating factors ranked third (MS 4.01, SD 1.121); clear incentive goal settings ranked fourth (MS 3.96, SD 1.046); safety incentive KPIs assessment procedures ranked fifth (MS 3.83, SD 1.046); incentive budget planning guidelines ranked sixth (MS 3.69, SD 1.261); guidelines for safety incentives design ranked seventh (MS 3.68, SD 1.215); establishing safety incentive determinants ranked eighth (MS 3.66, SD 1.220); long-term H&S development plan ranked ninth (MS 3.65, SD 1.260); proactive approach to safety management ranked tenth (MS 3.63, SD 1.298); fostering workplace positive attitudes ranked 11th (MS 3.63, SD 1.332); promoting workplace safety requirements ranked 12th (MS 3.61, SD

TABLE 8.2
Descriptive Analysis of Policy Drivers for Safety Incentives

Policy Drivers for Safety Incentives	Mean	Std. Deviation	Ranking
Enhance safety performance	4.04	1.054	1
Organisational workforce management	4.02	0.962	2
Provide interrelated motivating factors	4.01	1.121	3
Clear incentive goal settings	3.96	1.046	4
Safety incentive KPIs assessment procedures	3.83	1.151	5
Incentive budget planning guidelines	3.69	1.261	6
Guideline for safety incentives design	3.68	1.215	7
Establishing safety incentive determinants	3.66	1.220	8
Long-term H&S development plan	3.65	1.260	9
Proactive approach to safety management	3.63	1.298	10
Fostering workplace-positive attitudes	3.63	1.322	11
Promote workplace safety requirements	3.61	1.315	12
Organisational resource management	3.59	1.173	13
Support workers' behavioural change	3.08	1.192	14
Economy growth acceleration	3.07	1.192	15
Establish incentive selection approaches	3.07	1.237	16
Establish incentive implementation plans	3.00	1.248	17
Technology application in incentive mechanisms	2.94	1.233	18

TABLE 8.3
Test for Kruskal–Wallis Showing p-Values of Policy Drivers for Safety Incentives

Policy Drivers for Safety Incentives	*P*-values
Enhance safety performance	0.351
Organisational workforce management	0.097
Provide interrelated motivating factors	0.068
Clear incentive goal settings	0.917
Safety incentive KPIs assessment procedures	0.057
Incentive budget planning guidelines	0.930
Guideline for safety incentives design	0.064
Establishing safety incentive determinants	0.096
Long-term H&S development plan	0.549
Proactive approach to safety management	0.060
Fostering workplace-positive attitudes	0.296
Promote workplace safety requirements	0.069
Organisational resource management	0.098
Support workers' behavioural change	0.088
Economy growth acceleration	0.093
Establish incentive selection approaches	0.093
Establish incentive implementation plans	0.098
Technology application in incentive mechanisms	0.096

1.315); and organisational resource management ranked 13th (MS 3.63, SD 1.298). In addition, the remaining variables with medium significance are support workers' behavioural change, ranked 14th (MS 3.08, SD 1.192); economy growth acceleration, ranked 15th (MS 3.07, SD 1.192); established incentive selection approaches, ranked 16th (MS 3.07, SD 1.237); established incentive implementation plans, ranked 17th (MS 3.00, SD 1.248); and technology application in incentive mechanisms, ranked 18th (MS 3.00, SD 1.248).

Furthermore, the study employed the Kruskal–Wallis test to compare the difference in responses from construction professionals involved in the study. According to Yong and Pearce (2013) and Rea and Parker (2014), the Kruskal–Walis is a non-parametric test to determine the significant difference between three independent groups at a p-value at a 95% confidence level. The study findings suggest no statistically significant difference among respondents regarding the policy drivers of safety incentives implementation in the Nigerian construction industry as presented in Table 8.3.

8.7 DISCUSSION

The findings from the results ranked enhanced safety performance as a policy driver influencing safety incentive implementation in the construction industry. The Kruskal–Wallis test finding was also confirmed by indicating no statistically significant difference among participants' responses. The findings are consistent with the study of Fernández-Muiz et al. (2009) and Moosa and Oriet (2022), who discovered that the most significant goals of safety incentives are to enhance safety performance, increase employee performance, foster positive attitudes among workers, and increase job satisfaction. Safety incentives motivate workers to adhere to workplace safety behaviour, creating a safe working environment and aligning with organisational performance goals. They seek to provide proactive approaches to safety management and to provide needed training and support to workers for early detection and prevention of potential accidents (Adekunle et al., 2023; Ogundipe et al., 2024a). In addition, the study findings align with Workman (2018) and Adekunle et al. (2023), who noted that improving safety compliance on construction sites should be a top priority through proactive approaches to safety management. Respondents also identified that by understanding the drivers of the policy approach to safety incentives implementation, the knowledge of management of the organisational workforce using interrelated motivating factors to enhance workplace safety. In addition, Ward et al. (2016) and Ogundipe et al. (2024b) admitted that policy approaches balance many interrelated motivating factors essential for maximising personnel and resource management and establishing incentive selection approaches.

Similarly, the effective implementation of safety incentives hinged on clear incentive goal settings and incentive KPIs assessment procedures. This study's findings support Ogundipe et al. (2024b), who noted that these drivers provide a mix of policy strategies and instruments outlining basic rules, standard procedures, and templates for designing and implementing safety incentives within the construction industry. According to Kankaanpää (2010), Jiang et al. (2019), and Ogundipe et al. (2024b), a policy approach to safety incentives implementation provides an understanding of outlining measurable goals and performance evaluation to incentivise the construction workforce. As part of the construction firms' long-term H&S development plan, the drivers of the policy approach to implementing safety incentives provide strategies to understand guidelines for design, budget planning, and establishing safety incentive determinant factors within the construction industry. As Abubakari et al. (2018) noted, the study findings emphasise that the policy drivers of implementing incentive policy are essential for construction firms to plan, regulate, control, and guide their resources, growth, and development. Likewise, the study findings align with Howlett (2019), who established the need for construction firms to understand the drivers of implementing mixes of policy instruments guiding incentive budgets and accelerating economic growth.

Consequently, the study findings also established the drivers of the policy approach to implementing safety incentives to foster positive workplace attitudes among the construction workforce. The result corroborates the findings of Ahmed and Faheem (2021), who admitted that reward and penalty schemes foster positive workplace attitudes, align with performance-related goals, improve workplace safety practices, and share risk between clients and contractors. The result was aligned with previous studies' findings, which established the principles of moral hazard problems in contracts and stated that implementing bonus contracts provides individual stakeholders with incentives for meeting a particular performance threshold (Imhof &

Kräkel, 2014). Hence, safety incentives are designed and implemented using reward and penalty schemes; these help foster positive workplace attitudes among workers who often fail to comply with safe working conditions.

The policy approach also promotes workplace safety requirements among construction firms and stakeholders. This result supports the findings of Ahmed and Faheem (2021) and Ogundipe et al. (2024b), who affirmed that a policy approach to safety incentives implementation promotes workplace safety requirements to improve H&S planning, workplace safety behaviour, practices, and performance that often affect construction activities in developing countries. The workplace safety requirements will provide an understanding of H&S workplace conditions, employee training needs, the role of the safety committee and subcontractors, and specialised works and equipment safety, aligning with a standalone safety incentive policy.

However, other drivers such as workers' behavioural change, economic growth acceleration, incentive selection approaches, incentive implementation plans, and technology application in incentive mechanisms, although they can influence safety incentives implementation, are not considered by the study respondents as policy drivers influencing safety incentive implementation in the construction industry.

8.8 CONCLUSION AND RECOMMENDATIONS

The study examines drivers of policy approach to safety incentives implementation through construction professionals' perspectives to enrich its application among construction organisations in developing economies. This provides an understanding of using safety incentives as proactive approaches to improve H&S performance, foster positive attitudes among workers, and increase job satisfaction within the construction industry. The drivers were identified from a review of extant literature conducted before surveying construction professionals' perspectives. The study findings established drivers of the policy approach to safety incentives implementation in the construction industry to enhance safety performance. These include enhanced safety performance, organisational workforce management, interrelated motivating factors, clear incentive goal settings, safety incentive KPIs assessment procedures, incentive budget planning guidelines, guidelines for safety incentive design, establishing safety incentive determinants, a long-term H&S development plan, a proactive approach to safety management, fostering workplace-positive attitudes, promoting workplace safety requirements, and organisational resource management. Contrarily, drivers such as workers' behavioural change, economic growth acceleration, incentive selection approaches, incentive implementation plans, and technology application in incentive mechanisms are not considered significant. The study's practical implications inform that construction stakeholders should be guided by the significant policy drivers in implementing safety incentives in the construction industry in developing countries, particularly Nigeria. It also emphasised the importance of policy drivers to understand determinant factors of safety incentives and their implementation within the construction industry. It will also provide proactive measures to improve H&S practices and workers' motivation in the construction industry. Understanding these policy drivers would also stimulate workplace safety efficiency and strategies towards attaining SDGs goals 3 and 8 and Africa Agenda 2063 pillar 1 in providing a quality of life and well-being for the construction workforce. Conclusively, these findings could inform construction professionals and industry stakeholders of the importance of establishing a national safety incentive policy to take significant steps towards implementing safety incentives in the construction industry in developing countries. Hence, further studies could adopt and test the identified policy drivers in developing countries using different statistical tool analyses.

REFERENCES

Abubakari, M., Asamoah, P. K. B., & Agyemang, F. O. (2018). Ghana and sustainable development: The 40-year national development plan in retrospective. *Journal of Human Resource and Sustainability Studies*, *6*, 24–36. https://doi.org/10.4236/jhrss.2018.61024

Abubakar, M. A., Zailani, B. M., Abdullahi, M., & Auwal, A. M. (2022). Potential of adopting a resilient safety culture toward improving the safety performance of construction organisations in Nigeria. *Journal of Engineering, Design and Technology*, *20*(5), 1236–1256.

Abukhashabah, E., Summan, A., & Balkhyour, M. (2020). Occupational accidents and injuries in the construction industry in Jeddah city. *Saudi Journal of Biological Sciences*, *27* (8), 1993–1998.

Adekunle, S. A., Aigbavboa, C. O., Ejohwomu, O. A., Aliu, J., & Thwala, D. (2022). Leaders of tomorrow: Status and outlook of BIM implementation among young quantity surveyors. *Proceedings of International Structural Engineering and Construction*, *9*(1), 1–15.

Adekunle, S. A., Ikuabe, M., Aliu, J., Ogunbayo, B., Aigbavboa, C., Oyewo, W. O., et al. (2023). Understanding safety of construction sites: Construction site workers' experience. In *Emerging debates in the construction industry* (pp. 292–306). Routledge.

Afolabi, F. J., de Beer, P., & Haafkens, J. A. (2021). Can occupational safety and health problems be prevented or not? Exploring the perception of informal automobile artisans in Nigeria. *Safety Science*, *135*, 105097.

Agrasuta, V., & Nelson, A. (2013). *The adoption of green dentistry among dentists in Thailand.* Ministry of Public Health, Thailand. Available at: www.researchgate.net/profile/Voramon-Agrasuta/publication/281629128_The_Adoption_of_Green_Dentistry_among_Dentists_in_Thailand/links/55f0f6b208aef559dc46eb9d/The-Adoption-of-Green-Dentistry-among-Dentists-in-Thailand.pdf [Accessed 14 August 2022].

Ahmed, I., & Faheem, A. (2021). How effectively do safety incentives work? A randomised experimental investigation. *Safety and Health at Work*, *12*(1), 20–27. https://doi.org/10.1016/j.shaw.2020.08.001

Akuete, E., Nduka, D. O., & Ogundipe, K. E. (2022). Maintenance feedback mechanisms and limiting factors of post-occupancy evaluation in student residents of higher education institutions in Nigeria. *Journal of Construction in Developing Countries*, *27*(2), 17–41.

Alfandi, A. M., & Alkahsawneh, M. S. (2014). The role of the incentives and reward system in enhancing employee's performance "A Case of Jordanian Travel and Tourism Institutions". *International Journal of Academic Research in Business and Social Sciences*, *4*(4), 326.

Awwad, R., El Souki, O., & Jabbour, M. (2016). Construction safety practices and challenges in a Middle Eastern developing country. *Safety Science*, *83*, 1–11. https://doi.org/10.1016/j.ssci.2015.10.016

Bao, Q. L., Tran, S. V. T., Yang, J., Pedro, A., Pham, H. C., & Park, C. (2024). Token incentive framework for virtual-reality-based construction safety training. *Automation in Construction*, *158*, 105167.

Bauer, T., & Erdogan, B. (2012). *An introduction to organisational behavior* (V. 1.1). Flat World Knowledge Inc, Boston. Available at: https://2012books.lardbucket.org/pdfs/an-introduction-to-organisational-behavior-v1.1.pdf [Accessed 24 May 2023].

Boadu, E. F., Wang, C. C., & Sunindijo, R. Y. (2021). Challenges for occupational health and safety enforcement in the construction industry in Ghana. *Construction Economics and Building*, *21*(1), 1–21.

Bryson, A., Freeman, R., Lucifora, C., Pellizzari, M., & Perotin, V. (2011). *Pay for performance: Incentive pay schemes and employees' financial participation, draft paper*. European Science Foundation. Available at: www.frdb.org/wp-content/uploads/2020/07/XII-conf_Report21_2010NEW.pdf [Accessed 22 August 2022].

Choi, T. N., Chan, D. W., & Chan, A. P. (2012). Potential difficulties in applying the Pay for Safety Scheme (PFSS) in construction projects. *Accident Analysis and Prevention*, *48*, 145–155.

Dambisya, Y., & Dambisya, Y. (2007). *A review of non-financial incentives for health worker retention in East and Southern Africa* (Training and Research Support Centre. EQUINET Discussion Paper 44). University of Limpopo and EQUINET.

Del-Puerto, C. L., & Elliott, J. (2012). Cost benefits analysis of construction safety incentives programs. *AACE International Transaction*. 869.1–869.12.

DeSilva, S. (2004). *An introduction to performance and skilled-based pay systems*. Available at: www.ilo.org [Accessed 22 August 2022].

Enters, T., Durst, P., & Brown, C. (2003). What does it take? The role of incentives in forest plantation development in the Asia-Pacific region. *Unasylva*, *54*, 11–18.

Fernández-Muñiz, B., Montes-Peón, J. M., & Vázquez-Ordás, C. J. (2009). Relation between occupational safety management and firm performance. *Safety Science*, *47*(7), 980–991.

Field, A. (2009). *Discovering statistics using SPSS* (3rd ed.). Sage Publications.

Food and Agriculture Organisation of the United Nations (FAO). (1999). *Incentive systems for natural resource management* (Environmental Reports Series 2. FAO Investment Centre). Food and Agriculture Organisation of the United Nations. Available at: www.fao.org/3/x2247e/x2247e.pdf [Accessed 26 May 2023].

Ghana Statistical Service (GSS). (2016). *Ghana Statistical Service: 2015 Labour force report*. Available at: www2.statsghana.gov.gh/docfiles/publications/Labour_Force/LFS%20REPORT_fianl_21-3-17.pdf [Accessed 28 November 2022].

Ghasemi, F., Mohammadfam, I., Soltanian, A. R., Mahmoudi, S., & Zarei, E. (2015). Surprising incentive: An instrument for promoting the safety performance of construction employees. *Safety and Health at Work*, *6*(3), 227–232.

Hasan, A., & Jha, K. N. (2013). Safety incentive and penalty provisions in Indian construction projects and their impact on safety performance. *International Journal of Injury Control and Safety Promotion*, *20*(1), 3–12. https://doi.org/10.1080/17457300.2011.648676

Howlett, M. (2019). *Designing public policies: Principles and instruments*. Routledge.

ILO. (2006). *C187 – promotional framework for occupational safety and health convention, 2006* (No. 187). Available online: www.ilo.org/dyn/normlex/en/f?p=NORMLEXPUB:12100:0::NO::P12100_INSTRUMENT_ID:312332 [Accessed 17 August 2023].

Imhof, L., & Kräkel, M. (2014). Bonus pools and the informativeness principle. *European Economic Review*, *66*, 180–191.

Ji, L., Liu, W., & Zhang, Y. (2021). Research on the tournament incentive mechanism of the safety behaviour for construction workers: Considering multiple heterogeneities. *Frontiers in Psychology, 12*, 796295.

Jiang, W., Luo, L., Wu, Z., Fei, J., Antwi-Afari, M. F., & Yu, T. (2019). An investigation of the effectiveness of prefabrication incentive policies in China. *Sustainability*, 11(19), 5149.

Kalatpour, O., & Khavaji, S. (2016). Occupational injuries overview: A general descriptive study of the petrochemical construction industries. *Caspian Journal of Health Research, 2*(1), 37–43.

Kankaanpää, E. (2010). Economic incentives as a policy tool to promote safety and health at work. *Scandinavian Journal of Work, Environment & Health, 36*(4), 319–324.

Kiguhe, M. B. (2013). *The role of motivation in enhancing employees' performance in the public sector organizations in Tanzania: The case of immigration department* (Master dissertation, Mzumbe University).

Lee, B., & Kim, H. (2024). Evaluating the effects of safety incentives on worker safety behavior control through image-based activity classification. *Frontiers in Public Health, 12*, 1430697.

Liu, J., Wang, X., Nie, X., & Lu, R. (2022). Incentive mechanism of construction safety from the perspective of mutual benefit. *Buildings, 12*(5), 536.

Mahmoud, A. S., Ahmad, M. H., & Yatim, Y. M. (2020). Factors influencing management commitment to safety performance in the construction industry. In *Proceedings of international structural engineering and construction holistic overview of structural design and construction conference (Euro-Med-Sec 2020), 3–8 August 2020, Limassol, Cyprus.*

Martocchio, J. J. (2015). *Strategic compensation: A human resource management approach* (10th ed.). Pearson Education India.

Moda, H. M., Ofodile, N., Zailani, B. M., Abubakar, M. A., & Ibrahim, Y. M. (2023). Management support as a critical success factor (CSF) for changing worker's safety attitude: A case of the Nigerian construction industry. *International Journal of Construction Management, 23*(15), 2551–2557. https://doi.org/10.1080/15623599.2022.2077084

Moosa, M. H., & Oriet, L. P. (2022). Factors affecting safety performance in the construction industry: An empirical study using structural equation modelling. *International Journal of Occupational Safety and Ergonomics, 28*(2), 779–789.

Naderi, H., Shojaei, A., & Ly, R. (2023). Autonomous construction safety incentive mechanism using blockchain-enabled tokens and vision-based techniques. *Automation in Construction, 153*, 104959.

Ndizera, V., & Muzee, H. (2018). A critical review of Agenda 2063: Business as usual?. *African Journal of Political Science and International Relations, 12*(8), 142–154.

Ngacho, C., & Das, D. (2014). A performance evaluation framework of development projects: An empirical study of Constituency Development Fund (CDF) construction projects in Kenya. *International Journal of Project Management, 32*(3), 492–507.

Nwozor, A., Okidu, O., & Adedire, S. (2021). Agenda 2063 and the feasibility of sustainable development in Africa: Any silver bullet? *Journal of Black Studies, 52*(7), 688–715.

Ogunbayo, B. F., Ramabodu, M. S., Adewale, B. A., & Ogundipe, K. E. (2024) Strategies for successful monitoring and evaluation practices in construction projects. In *2024 international conference on science, engineering and business for driving sustainable development goals (SEB4SDG), Omu-Aran, Nigeria* (pp. 1–7). https://doi.org/10.1109/SEB4SDG60871.2024.10630137

Ogunde, A. O., Dafe, O. E., Akinola, G. A., Ogundipe, K. E., Oloke, O. C., Ademola, S. A., & Olaniran, H. F. (2017). Factors militating against prompt delivery of construction projects in Lagos megacity, Nigeria: Contractors' perspective. *Mediterranean Journal of Social Sciences, 8*(3), 233.

Ogundipe, K. E., Aigbavboa, C. O., & Ogunbayo, B. F. (2024a). Validating factors affecting a national safety incentives policy for the construction industry in developing countries. *Sustainable Construction in the Era of the Fourth Industrial Revolution, 149*, 50–61.

Ogundipe, K. E., Aigbavboa, C. O., & Ogunbayo, B. F. (2024b). A review of applicable approaches to safety incentive schemes design in the construction industry. In *2024 international conference on science, engineering and business for driving sustainable development goals (SEB4SDG), Omu-Aran, Nigeria* (pp. 1–8). https://doi.org/10.1109/SEB4SDG60871.2024.10630369

Ogundipe, K. E., Aigbavboa, C. O., & Ogunbayo, B. F. (2025). A mixed-method approach to validating barriers to safety incentives implementation in the Nigerian construction industry. *Frontiers in Built Environment, 11*, 1456893.

Ogundipe, K. E., Ogunbayo, B. F., & Aigbavboa, C. O. (2024). A review of barriers to safety incentives design and implementation in the construction industry. In *International conference on engineering, project, and production management* (pp. 459–470). Springer.

Ogundipe, K. E., Ogunde, A., Olaniran, H. F., Ajao, A. M., Ogunbayo, B. F., & Ogundipe, J. A. (2018). Missing gaps in safety education and practices: Academic perspectives. *International Journal of Civil Engineering and Technology (IJCIET), 9*(1), 273–289.

Ogundipe, K. E., Olanirah, F. H., Ogundipe, U. L., Ajao, A. M., & Ogunbayo, B. F. (2018). Assessing the impact of quality supervision on construction operatives' service delivery. *International Journal of Civil Engineering and Technology, 9*(9), 426–439.

Ogwueleka, A. C. (2015). *Evaluation of incentive mechanisms on performance-based contracting systems in South Africa and Nigeria* (Doctoral dissertation, University of Pretoria).

Oke, A. E., Aliu, J., Tunji-Olayeni, P., & Abayomi, T. (2024). Application of gamification for sustainable construction: An evaluation of the challenges. *Construction Innovation, 24*(4), 1066–1084. https://doi.org/10.1108/CI-09-2022-0247

Oni, O. Z., Olanrewaju, A., & Khor, S. C. (2023). Fuzzy synthetic evaluation of the factors affecting health and safety practices in Malaysia's construction industry. *Journal of Engineering, Design and Technology* (Ahead-of-print). https://doi.org/10.1108/JEDT-08-2022-0432

Oni, O. Z., Olanrewaju, A., & Khor, S. C. (2024). Identifying key accident causation factors in the Malaysian construction industry. *International Journal of Occupational Safety and Ergonomics, 30*(2), 366–377.

Opawole, A., & Jagboro, G. O. (2016). Benchmarking parties' obligations in the execution of concession-based PPP projects in Nigeria. *Journal of Place Management and Development, 9*(1), 27–46.

Osei-Asibey, D., Ayarkwa, J., Acheampong, A., Adinyira, E., & Amoah, P. (2021). An examination of causes of accidents and hazards in the Ghanaian construction industry. *Open Journal of Safety Science and Technology, 11*, 66–88. https://doi.org/10.4236/ojsst.2021.112006

Othman, I., Kamil, M., Sunindijo, R. Y., Alnsour, M., & Kineber, A. F. (2019). Critical success factors influencing construction safety program implementation in developing countries. In *Journal of Physics: Conference Series, Volume 1529, The 2nd Joint International Conference on Emerging Computing Technology and Sports (JICETS), 25–27 November 2019, Bandung, Indonesia, 1529*(4), 042079.

Owusu-Manu, D. G., Babon-Ayeng, P., Kissi, E., Edwards, D. J., Okyere-Antwi, D., & Elgohary, H. (2023). Green construction and environmental performance: An assessment framework. *Smart and Sustainable Built Environment, 12*(3), 565–583.

Pallant, J. (2016). *SPSS survival manual* (6th ed.). Open University Press.

Rea, L. M., & Parker, R. A. (2014). *Designing and conducting survey research: A comprehensive guide*. John Wiley & Sons.

Wang, Y., & Zhou, G. (2023). Analysis of incentive mechanism and contractor behaviours under informatisation construction in megaprojects. *Kybernetes, 53*(12), 5220–5241. https://doi.org/10.1108/K-04-2023-0696

Ward, S. C., Bagley, C., Lumby, J., Hamilton, T., Woods, P., & Roberts, A. (2016). What is 'policy' and what is 'policy response'? An illustrative study of the implementation of the Leadership Standards for Social Justice in Scotland. *Educational Management Administration & Leadership*, *44*(1), 43–56. https://doi.org/10.1177/1741143214558580

Wilde, G. J. S. (2011). Safety incentives programmes. In *Encyclopedia of occupational health and safety*. Available at: www.iloencyclopaedia.org/part-viii-12633/safety-programs/item/993-safety-incentive-programmes [Accessed 6 September 2022].

Workman, S. (2018). Are safety incentive programs effective? *Professional Safety*, *63*(2), 66–67.

Wu, Y., Schwebel, D. C., & Hu, G. (2018). Disparities in unintentional occupational injury mortality between high-income countries and low- and middle-income countries: 1990–2016. *International Journal of Environmental Research and Public Health*, *15*(10), 2296.

Yamane, T. (1967). *Statistics, an introductory analysis* (2nd ed.). Harper and Row.

Yong, A. G., & Pearce, S. (2013). A beginner's guide to factor analysis: Focusing on exploratory factor analysis. *Tutorials in Quantitative Methods for Psychology*. *9*(2), 79–94.

Zailani, B. M., Abubakar, M., & Ibrahim, Y. M. (2021, August). The role of safety attitude in changing safety behaviour and hazard recognition capability of construction workers. In *West Africa built environment research (WABER) conference, Accra, Ghana.*

Zhu, J., Zhang, C., Wang, S., Yuan, J., & Li, Q. (2022). Evolutionary game analysis of construction workers' unsafe behaviours based on incentive and punishment mechanisms. *Frontiers in Psychology*, *13*, 907382.

9 Ergonomics in Construction

Bridging Technological Gaps with a Unified BIM Framework for Agenda 2030

Mascha Samkova, Panagiotis Markopoulos, Georgios Markopoulos and Akash Nandi

9.1 INTRODUCTION

Since the introduction of the first Environmental, Social, and Governance (ESG) Index in 1991, sustainable, social, and ethical considerations in investment, corporate management, and governance have become not only a widespread global trend but also a crucial factor in evaluating companies' performance and risk management strategies. ESG principles have fundamentally reshaped the way investments are made, funding is allocated, public and market attention is garnered, acceptance and approvals are granted, projects are executed, and corporate success is measured. The widespread adoption of ESG criteria has led businesses to rethink their strategies and redefine their operational frameworks to align with these principles. However, for this transformation to be effectively implemented, it must be supported by appropriate technological advancements that enable efficient data collection, analysis, and utilization. Equally important is the establishment of a cohesive corporate culture that ensures alignment between business operations and ESG compliance, particularly in emerging markets. The extent to which corporate cultures in these regions embrace and integrate ESG standards plays a significant role in determining how successfully they can bridge the technological gap with more developed economies (Alkatheeri et al., 2023; Markopoulos et al., 2023).

Post-conflict regions serve as a prime example of where large-scale infrastructure projects must be aligned with ESG principles to foster long-term, sustainable reconstruction. These projects are often critical for rebuilding communities and restoring economic stability. One such example is the United Nations Development Programme's (UNDP) initiative to rebuild the Ngarannam village in northeast Nigeria's Borno State after the community was devastated by the Boko Haram insurgency in 2015 (UNDP, 2022). Located in the Chad Basin, the village reconstruction is adapted to the region's climate vulnerability. The village's housing includes

DOI: 10.1201/9781003672036-9

two water boreholes, greening projects, energy-efficient stoves, solar-powered street lights, and government offices. Successful post-conflict climate-mitigating reconstruction efforts require not only high financial and technical resources but also a well-defined and universally understood corporate culture that guides the adoption of sustainable construction methods with the ultimate end goal of providing human security of livelihoods. Without the shared mission to adopt human-centered design in construction work in line with the UN SDGs' framework, the efficiency and long-term viability of such projects could be compromised.

Beyond corporate culture, the necessity of a common language for ESG integration and technological adoption cannot be overstated. Effective coordination and communication are crucial to ensuring that ESG principles are embedded in corporate strategies and operational workflows. In some cases, such a shared language already exists and can be readily implemented across different sectors and regions.

However, in other cases, the absence of a unified terminology creates barriers to collaboration, requiring the development of a new, standardized framework for communication and integration. In the construction sector and the built environment worldwide, the importance of a unified language is particularly pronounced. The industry faces several long-standing challenges, including low productivity levels, prolonged project timelines, and costs that frequently exceed initial estimates.

These inefficiencies are especially prevalent in emerging markets, where reliance on outdated methodologies and a slow adoption rate of advanced technologies contribute to continued underperformance. Moreover, rising material costs in global markets further exacerbate these challenges, placing additional financial strain on construction projects.

At the same time, external pressures are driving the industry toward more sustainable construction practices. The United Nations Sustainable Development Goals (UN SDGs) have established a global framework that encourages businesses and industries to prioritize environmental, social, and economic sustainability in their operations. In particular, initiatives like the UN Global Compact which is the world's largest corporate social responsibility (CSR) initiative, comprising 20,000 businesses across 90% of UN member states, are playing a crucial role in accelerating the transition to sustainable construction practices. Under the Agenda 2030 framework, companies are in the next five years increasingly incentivized to incorporate sustainable materials, adopt innovative building techniques, and improve their overall ESG performance. Higher ESG scores not only contribute to long-term business viability but also enhance a company's reputation and investment appeal (Markopoulos et al., 2023).

In this evolving landscape, the integration of ESG principles, corporate cultural alignment, and technological advancement will be critical in shaping the future of sustainable construction. The development of standardized methodologies, industry-wide best practices, and a common language for ESG compliance will play an essential role in ensuring that the construction sector effectively meets the demands of sustainability, efficiency, and global competitiveness.

9.2 CONSTRUCTING THE LONG-TERM IMPACT OF ESG CRITERIA AND UN SDGS

The reputation of a company is increasingly shaped by its alignment with global sustainability frameworks, particularly the UN SDGs and the ESG Index. These frameworks have become essential markers of corporate responsibility and sustainability, significantly influencing a company's brand perception, market positioning, and financial performance (Jeffrey et al., 2019). Research has consistently demonstrated a strong and positive correlation between ESG performance and corporate reputation. Companies that maintain stable and high ESG ratings tend to experience enhanced public trust, increased consumer loyalty, and stronger investor confidence, leading to long-term business advantages (Meng et al., 2023).

As investor awareness and scrutiny of corporate ESG strategies continue to rise, companies with strong and transparent sustainability practices benefit from greater financial support and strategic partnerships. A study has shown that the presence of CSR initiatives, as measured by content analysis of CEO social media engagement (particularly tweets about the SDGs) can contribute significantly to an improved corporate reputation (Grover et al., 2019). This, in turn, attracts responsible investors who prioritize ethical and sustainable business practices when making investment decisions (Hadro et al., 2022). The growing emphasis on ESG and sustainability is reflected in global investment trends. In 2019, sustainable investment surpassed $30 trillion, representing a 68% increase since 2014 and a tenfold increase compared to 2004. This significant rise in ESG-focused investments over the past two decades has been driven by heightened social, governmental, and consumer concerns regarding the environmental and societal impact of corporate activities. Companies that proactively integrate ESG considerations into their business models and recognize the long-term value of sustainability are better positioned to thrive in the evolving economic landscape (Koller et al., 2019).

The construction sector plays a crucial role in advancing ESG goals due to its direct impact on infrastructure development, urbanization, and resource consumption. Buildings are fundamental to a nation's economic and social infrastructure, serving as essential spaces for living, working, and social interaction. Consequently, construction companies are uniquely positioned to incorporate human-centric designs that prioritize sustainability, energy efficiency, and environmental stewardship.

As these companies fulfill critical economic and functional roles, they are also expected to adopt innovative approaches that align with global sustainability targets.

From a strategic perspective, the ESG framework and the UN SDGs serve as unifying principles that encourage companies across industries to develop innovative strategies aimed at maximizing public good while ensuring financial viability (Markopoulos & Markopoulos, 2023). Organizations are increasingly aligning their corporate strategies with specific UN goals, such as SDG 11 (Sustainable Cities and Communities) and SDG 12 (Responsible Consumption and Production) (Visvizi & del Hoyo, 2021). By integrating these objectives into their business models, companies can systematically assess their ESG impact, conduct comprehensive risk assessments to mitigate external threats and internal vulnerabilities, and pioneer new construction methodologies that enhance efficiency, resilience, and sustainability (LePage & Renaerts, 2023).

From an operational and process-driven perspective, the construction industry is progressively adopting sustainable practices and integrating ESG principles into its core business functions. This shift has facilitated the achievement of multiple SDGs through various measures, including the use of eco-friendly and recyclable materials, the implementation of innovative building techniques, the enhancement of energy efficiency, and the reduction of water consumption and waste generation (Grossi & Trunova, 2021). These sustainability-focused advancements contribute to minimizing the industry's environmental footprint while promoting long-term economic and social benefits.

Furthermore, the construction sector plays a vital role in fostering improved working conditions and overall well-being, contributing directly to SDG 3 (Good Health and Well-Being) (Opoku, 2022). Sustainable construction practices extend beyond physical structures to influence broader supply chain management, encouraging ethical sourcing of materials, fair labor practices, and the adoption of environmentally responsible procurement processes. The integration of renewable energy solutions in new construction projects supports SDG 7 (Affordable and Clean Energy), while efforts to provide affordable housing, enhance indoor air quality, optimize natural lighting, and incorporate green spaces contribute to SDG 11's objectives for sustainable urban development.

As the construction industry continues to evolve, the incorporation of ESG and SDG principles is expected to drive further innovation, resilience, and long-term economic growth. Companies that actively engage in sustainability-driven strategies are not only contributing to global development goals but also ensuring their own competitive advantage in an increasingly environmentally conscious market. By leveraging ESG as a guiding framework, the construction sector has the potential to transform urban environments, create more livable and inclusive communities, and establish a model for responsible and sustainable business practices.

9.3 BUILDING INFORMATION MODELING IN CONSTRUCTING SUSTAINABILITY

Building information modeling (BIM) has evolved over the past two decades into a cornerstone of sustainable construction, reshaping the way buildings are designed, constructed, and maintained. As a revolutionary methodology in the construction industry, BIM facilitates the integration of advanced digital technologies to enhance efficiency, reduce waste, and optimize resource utilization. In particular, BIM plays a critical role in the sustainable construction movement by promoting the use of recycled materials sourced from the circular economy, minimizing energy consumption, and managing variable costs associated with operational efficiency. Additionally, BIM serves as a powerful tool for monitoring and controlling the projected carbon emissions of buildings, helping to mitigate their long-term environmental impact (Deng et al., 2021).

Sustainable construction, as defined by Jackson (2021), is a practice centered around the responsible and efficient use of natural resources while designing healthier, more energy-efficient spaces. This approach ensures that architectural blueprints take into account a building's future ecological footprint, emphasizing the importance of environmentally conscious design from the early planning stages. BIM is particularly aligned with the principles of sustainability, as it allows for the seamless storage and management of vast amounts of data related to construction projects, including energy consumption patterns, material usage, and carbon footprints. Furthermore, BIM provides comprehensive workflow data throughout construction operations, enabling project managers, engineers, and stakeholders to make well-informed, data-driven decisions that support sustainable development (Najjar, 2017).

To fully appreciate the transformative impact of BIM in sustainable construction, it is essential to explore the key technologies and software solutions associated with this methodology. Various digital tools and AI-powered applications have been developed to enhance the design and execution of construction projects, streamlining processes while improving accuracy and efficiency. Among the most notable BIM technologies are Autodesk Revit, Graphisoft ArchiCAD, and Solibri, all of which integrate artificial intelligence (AI) to optimize architectural design and structural modeling (Waas, 2022). A key feature of these platforms is the use of generative design, an advanced computational approach that enables architects and engineers to generate multiple design solutions based on predefined constraints and project-specific goals. By leveraging AI, these tools not only enhance design precision but also reduce repetitive and time-consuming manual tasks, allowing for a more streamlined and efficient design process.

Beyond AI-enhanced design tools, virtual reality (VR) and augmented reality (AR) technologies have emerged as powerful assets in the BIM ecosystem, particularly in the context of digital twin cities and metaverse-based construction planning. Such technologies can be integrated with gamification mechanics to increase user engagement and performance efficiency (Markopoulos et al., 2022). A wide range of VR and AR applications are now being employed to enhance collaboration and visualization in architectural design and construction projects. These include VR collaboration tools such as IrisVR and Varjo, real-time rendering and VR solutions like Enscape, and AR-based construction tools such as Microsoft HoloLens and ARki. Additionally, smart helmets, such as the Daqri Smart Helmet, offer AR capabilities to construction professionals, improving onsite safety and operational efficiency. Unity Reflect, a cutting-edge real-time 3D and AR/VR tool, further enhances the ability of construction teams to interact with and refine BIM models. The integration of these technologies allows project stakeholders to immerse themselves in highly detailed 3D models, providing an interactive and immersive experience that improves comprehension, facilitates decision-making, and enhances overall project coordination. The benefits of VR and AR in the construction industry extend beyond design visualization, as these technologies also support enhanced accuracy, improved collaboration, optimized training methods, and increased safety for onsite workers.

In addition to VR, AR, and AI-driven design platforms, several other advanced BIM technologies are contributing to the sustainable evolution of the construction industry. For example, drones and photogrammetry tools, such as Trimble Tekla, are increasingly being used for robotics-integrated modeling, enabling high-resolution site analysis and progress monitoring. The integration of the Internet of Things (IoT) with BIM has also revolutionized building performance monitoring, allowing real-time data collection and analysis to optimize energy efficiency, structural integrity, and environmental impact. Other cutting-edge BIM technologies include laser scanning, digital fabrication, and assembly solutions like Asmbld, which provide precise modeling capabilities to ensure sustainable and cost-effective construction practices. The use of point cloud data and digital twin technology, exemplified by Bentley Systems, further enhances real-time analytics by leveraging IoT-driven insights to bridge the gap between office-based models and onsite construction activities.

A closely related technology that complements BIM is 3D modeling and 3D printing, which have revolutionized the way building components are designed and fabricated. 3D printing, in particular, has emerged as a game-changer in sustainable construction by allowing for the direct printing of architectural blueprints and even entire structural components using recycled or bio-based materials.

This approach not only reduces material waste but also enhances the speed and efficiency of the construction process. Additionally, the development of new software platforms, such as Sefaira, has introduced advanced tools specifically designed for monitoring the real-time sustainable performance of buildings. Sefaira provides critical insights into material usage, construction practices, and environmental impacts, making it an increasingly valuable asset for sustainable construction initiatives.

On the frontier of machine learning and data analytics, BIM technologies continue to push the boundaries of efficiency, collaboration, and risk mitigation in construction. Cloud computing solutions, such as Autodesk BIM 360, facilitate seamless collaboration between design teams, contractors, and project managers, ensuring that all stakeholders have access to up-to-date project data. Meanwhile, machine learning algorithms are being employed for predictive analysis, enabling construction teams to anticipate potential design conflicts, material shortages, and operational risks before they occur. Blockchain technology is also being explored as a means of enhancing BIM data security and ensuring the integrity of digital construction records. Additionally, generative design tools like Autodesk Dynamo leverage AI-driven algorithms to optimize building layouts, material selection, and energy performance.

Overall, the adoption of BIM in sustainable construction has led to a significant reduction in environmental impacts while promoting both social and economic development. By integrating digital technologies, data analytics, and AI-driven design strategies, BIM has transformed the way construction projects are planned, executed, and maintained (Díaz & Antón, 2014). One of the most critical advantages of BIM is its ability to facilitate seamless collaboration among architects, engineers, contractors, and project owners, ensuring that all stakeholders have access to essential project information and analytics. This enhanced level of collaboration enables more informed decision-making at the executive level, reducing inefficiencies and improving project outcomes.

Furthermore, BIM's ability to simulate different design scenarios allows architects and engineers to explore various energy-efficient solutions, optimizing building performance while minimizing long-term environmental impacts. This, in turn, contributes to significant reductions in future energy consumption and carbon emissions, ensuring that buildings remain sustainable throughout their entire life cycle. The integration of BIM into the certification processes for green building standards, such as LEED (Leadership in Energy and Environmental Design) and BREEAM (Building Research Establishment Environmental Assessment Method), further underscores its importance in achieving environmental compliance and promoting sustainable construction practices worldwide.

As regulatory requirements and industry standards continue to evolve, the role of BIM in sustainable construction will become even more pivotal. By leveraging BIM technologies, construction professionals can align their projects with global sustainability goals, including the UN SDGs and ESG criteria, thereby contributing to the broader mission of creating resilient, energy-efficient, and environmentally responsible built environments (Markopoulos & Ramonda, 2022). With ongoing advancements in AI, machine learning, and digital twin technology, the future of BIM is poised to drive further innovation, enhancing the construction industry's capacity to address the challenges of climate change, resource scarcity, and urbanization. Ultimately, BIM represents a transformative force that will continue to shape the future of construction, paving the way for more sustainable, efficient, and technologically advanced building practices.

9.4 UNIFYING BIM TOOLS TO BUILD UP COMPLIANCE WITH ESG AND SDGS

The application of BIM technology in real-world projects presents multiple challenges, two of the most significant being: (1) determining the minimum technological architecture necessary for BIM to be considered sufficiently integrated into an organization's operations and sustainability efforts, and (2) fostering an organizational culture that promotes digital information creation and seamless knowledge-sharing among all stakeholders involved in a project.

Successfully addressing these challenges is critical for maximizing the benefits of BIM technology in the construction and infrastructure industries.

The first challenge revolves around the need to establish a baseline technological architecture that ensures BIM contributes meaningfully to an organization's workflow and sustainability objectives. This requires a careful selection of BIM technologies, tailored to the specific needs of each project and client. Without a strategic approach to implementation, organizations risk overcomplicating their systems, leading to excessive costs associated with software, hardware, and specialized expertise. Overly complex implementations may create resistance among professionals, ultimately undermining the effectiveness of BIM adoption.

One of the primary reasons BIM is still only partially developed and applied across the industry is the hesitancy to invest in costly and intricate digital solutions (Liao et al., 2022). Many companies face financial constraints or lack the necessary expertise to fully exploit the advantages of BIM, leading to selective or fragmented adoption. A more agile and adaptive approach to BIM integration is required. One that prioritizes the essential technological components necessary for efficient project execution while avoiding unnecessary financial and operational burdens.

The second challenge involves fostering an organizational culture that supports digital collaboration and seamless information sharing. Beyond technological considerations, the successful adoption of BIM also depends on the development of a strong organizational culture that encourages digital information creation, collaboration, and knowledge-sharing among all stakeholders. A key barrier to full BIM implementation is the reluctance of project participants to openly share data due to concerns about intellectual property, competitive advantage, and lack of trust between different parties involved in a project.

To overcome these cultural barriers, organizations must establish a collective goal that aligns the interests of all stakeholders and incentivizes collaboration. One potential strategy is to integrate BIM technology with widely recognized sustainability frameworks, such as the UN SDGs and the ESG Index (Markopoulos, Markopoulos, Nandi, Wu, et al., 2024). While the UN SDGs focus on achieving long-term social and environmental sustainability, the ESG framework extends these objectives by incorporating financial and economic sustainability considerations. Aligning BIM adoption with these sustainability goals provides a strong incentive for organizations to improve digital collaboration, as it directly contributes to long-term environmental, social, and financial benefits for all project participants.

By leveraging BIM in conjunction with SDG and ESG principles, organizations can create a more structured and goal-oriented approach to digital transformation. This alignment not only enhances efficiency and transparency but also promotes responsible business practices that are essential for long-term industry growth. Figure 9.1 illustrates how the combination of BIM Agile methodologies and sustainability-driven frameworks can significantly enhance BIM productivity, ensuring that both

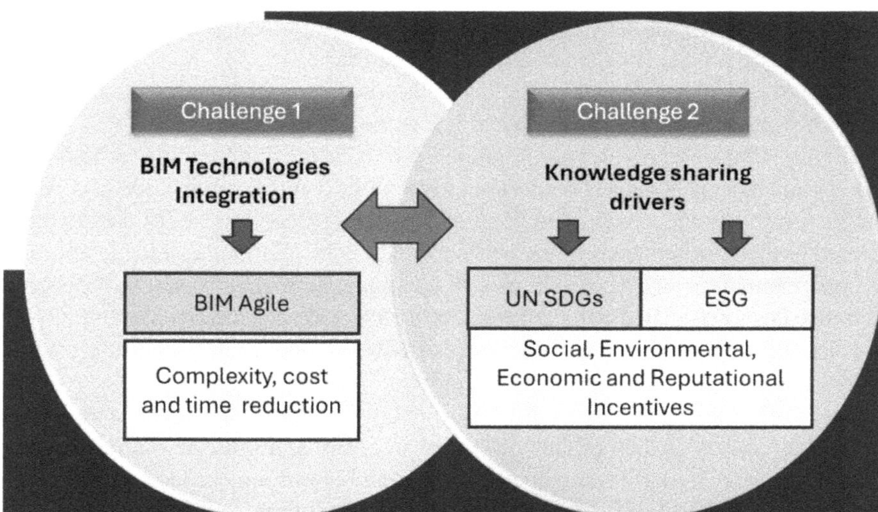

FIGURE 9.1 Building Information Modeling (BIM) challenges and resolution approaches.

technology and culture work in harmony to drive innovation and efficiency in the built environment.

Table 9.1 provides an initial attempt to address both challenges by integrating key BIM technologies with sustainability frameworks. In the first column, the table identifies essential BIM technologies that are considered critical to the success of construction projects, highlighting their significant contributions to project outcomes.

The second column correlates each BIM technology with specific UN SDGs, illustrating how the use of these technologies aligns with global sustainability efforts. This connection helps demonstrate how BIM can support broader environmental and social objectives (Jing & Alias, 2024).

In the third column, the table links each BIM technology to the relevant ESG Pillar categories and themes, based on the Refinitiv ESG scoring methodology (LSEG, 2024). This categorization allows for a clear understanding of how BIM can contribute to a company's sustainability and governance practices, in line with internationally recognized ESG standards.

The analysis presented here serves as a preliminary overview, offering a foundation for further exploration. It can be expanded by incorporating additional direct or indirect BIM technologies that are currently being utilized in the industry, as mentioned in previous examples. Additionally, the analysis could be developed further by delving deeper into the 169 SDG targets and **234** unique indicators outlined in the global indicator framework. This approach would allow for a more nuanced understanding of how BIM technologies can contribute to achieving specific sustainability targets.

Moreover, the analysis can be structured to consider the three core dimensions of sustainable development (economic, environmental, and social) providing a more detailed and comprehensive assessment of BIM's role in advancing sustainable construction practices.

The combinations of selected BIM technologies presented in Table 9.1 are closely linked to both the internal and external strategic objectives of the company. These corporate goals can encompass a variety of factors, such as implementation costs, project timelines, labor requirements, availability of expertise, human resources, team dynamics, local conditions, and other relevant variables. The chosen technologies inherently influence the social and sustainability outcomes of the project and its execution process.

For instance, the integration of VR technology within BIM contributes to Goal 13 (Climate Action) by promoting energy conservation and reducing transportation time, thus supporting environmental sustainability. Additionally, VR technology helps advance Goal 10 (Reduced Inequalities) by making projects more accessible to individuals with disabilities, fostering greater inclusivity within the construction process (Dulau et al., 2019).

These same VR technologies are also aligned with various ESG themes, including Emission (E), Environmental Management Systems (E), and Energy (E), reflecting their positive environmental impact. In the social domain, VR contributes to Diversity and Inclusion (S) and Human Rights (S), while in governance, it enhances the Management Structure (G).

TABLE 9.1
BIM technologies alignment with UN SDGs and the ESG criteria

BIM Technology	Key UN SDGs (Number)	Key ESG Pillar (E, S, or G); Theme
Autodesk Revit	Industry Innovation and Infrastructure (9)	Product Innovation (E), Product Quality (S), Career Development and Training (S)
Graphisoft ArchiCAD (AI-powered)	Industry Innovation and Infrastructure (9)	Product Innovation (E), Product Quality (S)
VR Collaboration Tools	Climate Action (13), Reduced Inequalities (10), Gender Equality (5)	Emission (E), Environmental Management Systems (E), Energy (E), Diversity and Inclusion (S), Human Rights (S), Management Structure (G)
AR for Construction	Sustainable Cities and Communities (11), (9)	Product Innovation (E), Product Quality (S)
Drones and Photogrammetry	Decent Work and Economic Growth (8), Responsible Consumption and Production (12)	Health and Safety (S), Product Innovation (E), Product Quality (S), Working Conditions (S)
Digital Twins	Gender Equality (5), (10), (8), (9), (12)	Health and Safety (S), Product Innovation (E), Product Quality (S), Diversity and Inclusion (S)
Cloud Computing	Industry Innovation and Infrastructure (9), (8)	Product Innovation (E), Product Quality (S), Career Development and Training (S), Data Privacy (S)
Machine Learning for Predictive Analysis	Industry Innovation and Infrastructure (9), (8), (12)	Product Innovation (E), Product Quality (S), Career Development and Training (S)
3D Printing	Good Health and Well-Being (3), (8), (9), (11), (12)	Health and Safety (S), Product Innovation (E), Product Quality (S), Working Conditions (S)
IoT for Building Performance Monitoring	(8), (9), (11)	ESG Reporting Transparency (G), Product Innovation (E), Product Quality (S), Career Development and Training (S)
Blockchain for BIM Data Security	(8), (9), (11), (12)	ESG Reporting Transparency (G), Product Innovation (E), Product Quality (S), Career Development and Training (S), Data Privacy (S)
Digital Fabrication and Assembly	(5), (10), (8), (9), (11), (12)	Working Conditions (S), Product Innovation (E), Product Quality (S), Career Development and Training (S)

This integration of VR technology underscores its broad applicability across multiple sustainability frameworks, highlighting its transformative potential in both project implementation and broader organizational strategy.

9.5 FRAMEWORK OF UBIML-C

The integration and alignment of BIM technologies with ESG standards, as well as the UN SDGs, form the foundation for what is referred to as a Unified Building Information Modeling Language (UBIML). This approach must be tailored and applied according to national contexts and the specific demands of each organization to ensure effective communication and coordination among stakeholders.

The political, corporate, and societal conditions under which a project operates can vary significantly, influencing the adoption and application of BIM technologies. These conditions are not limited to business operations alone but extend to a wide range of factors, including cultural norms, religion, social structures, climate, geography, legislation, financial institutions, corruption, and others. For example, a project in a country with strict religious and social norms—such as those that restrict women from working on construction sites or engaging in physically demanding tasks—might require the use of enhanced VR and AI technologies to support remote participation and facilitate design and project management tasks without breaching these societal norms (Markopoulos & Luimula, 2020). Drones, on the other hand, could be particularly useful in regions with difficult-to-reach terrain, where traditional ground transportation may not be feasible. Similarly, Blockchain technology could serve as an alternative for secure financial transactions and monitoring in regions with unstable economies or weak financial institutions, particularly in emerging or crisis-driven economies with low credit ratings. Conversely, technologies like Cloud Computing or Digital Twins might not provide significant societal value in areas where communication infrastructure is unreliable or underdeveloped (Markopoulos, Markopoulos, Nandi, Zhao, et al., 2024).

Moreover, the external factors impacting these technologies may also be influenced by the goals outlined in the UN SDGs and ESG frameworks. While many ethical construction companies around the world operate in alignment with these frameworks, not all of them view the SDGs and ESG standards as mandatory practices, particularly in regions where local government incentives or cost-efficiency are lacking. However, this does not imply that these companies are environmentally harmful, socially irresponsible, or unethical. Rather, they operate according to their own set of cultural and societal norms, which may not always align with global sustainability frameworks but are still consistent with their local context.

In this context, the conditions specific to a region or organization form the semantics of the UBIML, while the technologies themselves represent the words in that language. The goals, whether they are the UN SDGs or ESG criteria, form the culture of the language. By understanding the interplay between these three elements—conditions, technologies, and goals—and effectively communicating the appropriate objectives, a common and adaptable language can be established that

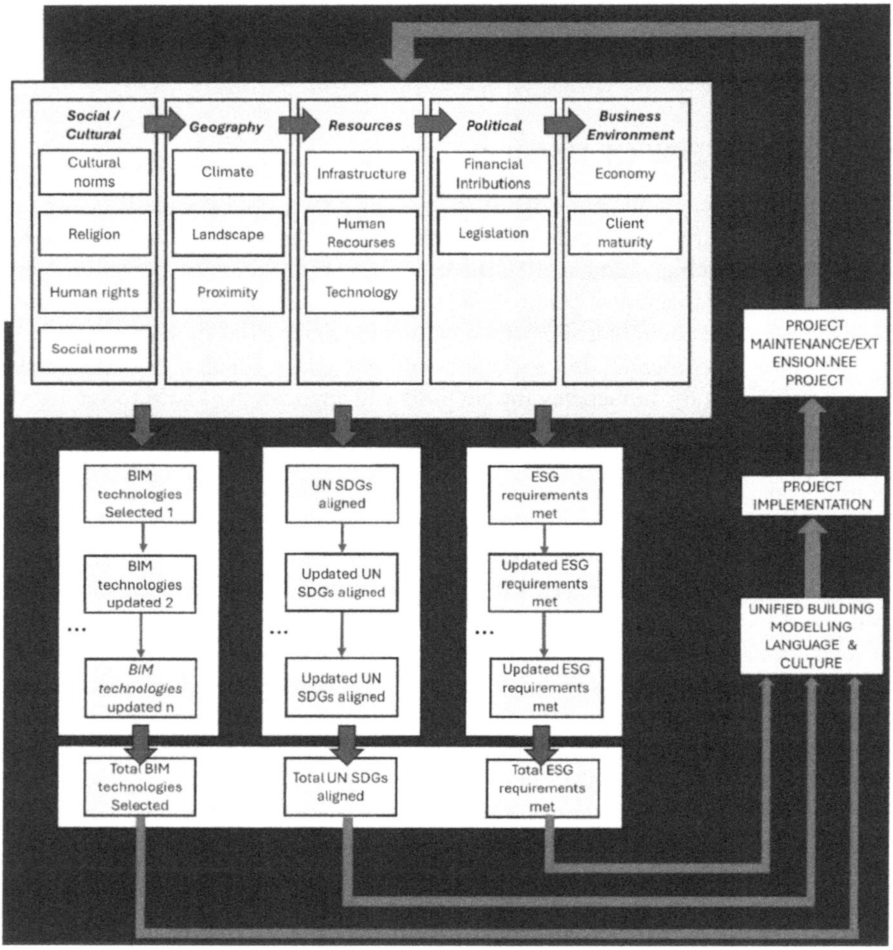

FIGURE 9.2 UBIML-C generator architecture.

enables stakeholders from diverse backgrounds to collaborate on a shared project, regardless of geographic location or cultural differences.

Figure 9.2 illustrates a high-level framework for generating a UBIML-C for each specific project. This framework operates across five distinct categories of conditions, each of which contributes to the selection of appropriate BIM technologies, alignment with the UN SDGs, and adherence to ESG requirements. As the categories are completed, the BIM technologies, SDGs alignment, and ESG criteria are updated to reflect the specific needs and requirements of the project. Upon completion of this process, the final set of technologies is determined, the SDGs are aligned with the project, and the relevant ESG requirements are identified. These results culminate in the creation of a UBIML-C, a comprehensive guide for the execution of the project.

In the case of project extensions, maintenance, or updates requested by the construction company due to new or evolving requirements, the same process is repeated, ensuring that any changes or additional needs are addressed while maintaining alignment with the project's sustainability goals and operational needs.

9.6 ORGANIZATIONAL CULTURE PRE- AND POST-CONDITIONS OF UBIML ADAPTATION

The UBIML has the potential to serve as a crucial tool for establishing a common understanding and aligning expectations when adopting BIM on various projects, especially those with an international scope. In his book *The Flat World*, Thomas Friedman makes several key assumptions about the nature of a globally interconnected world. He argues that advancements in technology and the deregulation of markets have created a world where businesses can conduct projects and operations from virtually anywhere, at any time, and by anyone (Friedman, 2007). According to Friedman, the rapid integration of economies and the opening of markets have led to an acceleration in economic processes, reducing the geographical and temporal distances between different stages of the supply chain. This shift has significantly altered the way businesses operate, allowing for the smoother flow of goods and services.

However, Pankaj Ghemawat, in his work *Distance Still Matters*, challenges the notion that the world has become entirely "flat", suggesting that the reality is more complex. Ghemawat argues that despite the forces of globalization, the world is still marked by significant cultural, administrative, geographic, and economic distances that complicate the idea of a completely borderless economy [Ghemawat, 2021]. These "distances" or barriers present obstacles that prevent generalized assumptions about the way global business operates, making the world "bumpy" rather than entirely smooth (Ramnath, 2011). Ghemawat's perspective highlights the fact that despite technological advancements and interconnectedness, there are still factors that shape how projects are conducted, especially when it comes to large-scale international collaborations.

Given these real-world complexities, certain pre- and post-conditions must be met to effectively create and implement the appropriate UBIML-C for a given project. The pre-conditions for successfully utilizing the UBIML-C include the existence of adequate BIM expertise within the project team, as well as the development of a suitable regulatory framework that governs the construction project.

Furthermore, the client's maturity in adopting BIM must be considered, particularly their understanding of the time, effort, and cost investments required for a BIM-driven project. Educating the client about the long-term benefits of BIM and ensuring that incentives are in place for aligning the project with UN SDGs and ESG reporting requirements are also crucial. In countries where cost-cutting, speed, and efficiency often dominate project priorities, the UBIML-C can be particularly valuable. It provides a structured way to implement at least the minimum necessary BIM processes in such environments, where quality and reliability are often treated as cost centers rather than potential drivers of profit.

The post-conditions for the application of UBIML-C relate to the effective utilization and dissemination of the benefits, on the other hand, associated with this unified approach to BIM. Promoting the advantages of adopting UBIML-C can provide substantial reputational benefits at local, regional, and international levels. This positive reputation, built on the alignment with ESG standards and SDGs, can drive broader adoption of the UBIML-C in other construction projects, particularly in competing companies. As companies increasingly recognize the value of sustainable development and corporate responsibility, these reputational benefits are likely to become a key factor in shaping future business strategies. Over time, the widespread recognition of the long-term benefits of BIM, as it is unified with UN SDGs and ESG frameworks, may lead to the gradual replacement of traditional BIM practices with more sustainable and culturally aligned approaches. This shift will ensure that sustainability is not just a buzzword but a core element of every construction project, with the UBIML-C serving as the central tool for translating these values into actionable, globally recognized standards.

9.7 LIMITATIONS AND AREAS OF FURTHER RESEARCH

The research presented in this chapter has focused on identifying the evolving global circumstances, variables, and both the pre- and post-conditions necessary for the creation of a Unified Building Information Modeling Language and Culture (UBIML-C). This includes an exploration of the overall concept behind the UBIML-C, the need for such a language, and the framework that would guide its composition. While the development of the holistic framework represents a significant step toward understanding how a UBIML-C could function in practice, the research's practical dimension has certain limitations. Notably, the UBIML-C has not yet been fully tested or evaluated to assess its tangible value or impact compared to traditional BIM systems.

Given this, it is essential to conduct further research to bridge this gap and investigate the real-world applicability and effectiveness of the UBIML-C. One key direction for future research will be to develop a prototype that can generate a UBIML-C for each specific project case. This prototype will be built upon the expert system-type approach introduced in this chapter, providing an explanatory logic that details why the output is generated in a particular way. By focusing on the logic behind each decision and output, the research aims to offer transparency into how the UBIML-C adapts to varying project needs, conditions, and cultural contexts.

The role of expert systems in this context is particularly noteworthy, as they have long been recognized as tools capable of solving complex problems by mimicking the decision-making abilities of human experts. Originally developed in the 1980s, expert systems are now experiencing a resurgence in the form of explanatory artificial intelligence (XAI), which focuses on providing human-centered explanations for the decisions and outputs generated by AI systems (Sovrano & Vitali, 2024). The integration of XAI into the development of the UBIML-C prototype will enable project teams to better understand the rationale behind the selection of specific BIM technologies, UN SDG alignments, and ESG requirements. By offering clear, understandable explanations for the decisions made by the system, the prototype will foster greater transparency, trust, and confidence in the implementation of BIM technologies and their alignment with sustainable development goals.

This research, while laying the groundwork for the development of UBIML-C, is just the beginning.

The next steps will involve refining the prototype and conducting case studies to evaluate the practical benefits of the UBIML-C in comparison to traditional BIM approaches. These evaluations will aim to provide a clearer understanding of how this unified language and culture can drive greater sustainability, collaboration, and efficiency in the construction industry, particularly in international and cross-cultural project environments.

9.8 CONCLUSION

Since its introduction more than two decades ago, BIM has evolved into a comprehensive suite of digital technologies. Over this time, organizations have been working to identify the most effective way to adapt BIM in a manner that maximizes its benefits while also managing and controlling the associated costs. The challenge has always been to find a balance between realizing the significant potential of BIM and ensuring that its adoption is both financially feasible and operationally effective.

On the one hand, there is increasing industrial pressure to accelerate the adoption of BIM, especially in light of the United Nations 2030 Agenda for Sustainable Development. This global initiative provides a five-year window for organizations to integrate BIM technologies more widely and to enhance their operations in ways that contribute positively to the achievement of the UN SDGs. As this deadline approaches, the push is on to utilize BIM in ways that not only improve the efficiency and productivity of construction projects but also support the realization of global sustainability targets, leading to improved ESG ratings for companies across the construction sector. The urgency for adopting BIM practices in this context is heightened by the expectation that these practices will play a pivotal role in achieving the SDGs and improving sustainability performance.

There are numerous challenges, on the other hand, associated with evaluating the feasibility of BIM adoption, especially when it comes to large-scale international projects. As global supply chains become more complex and projects extend across multiple countries and types of economies, the costs associated with implementing BIM can rise significantly, sometimes making it difficult for organizations to evaluate whether BIM is the right choice in terms of cost-effectiveness and return on investment. These challenges are further complicated by the need to navigate a broad spectrum of variables, such as differing legal frameworks, cultural norms, and technological capabilities across different countries, all of which can impact the efficiency and success of BIM implementation (Migilinskasa et al., 2013).

In response to these complexities, this chapter has introduced the concept of the UBIML-C framework, a model designed to bridge the gap between BIM technologies, UN SDGs, and ESG performance. The UBIML-C framework seeks to provide a comprehensive approach to BIM implementation by integrating these three components in a way that allows organizations to predict the long-term impact of their strategic decisions, both in terms of their financial performance and their reputational standing, on both a local and international level.

The central idea behind UBIML-C is that in a globally interconnected world, the strategies and decisions related to project implementation cannot be solely driven by financial or technological considerations. Instead, they must also take into account the cultural norms and governance structures that exist both within national borders and across international contexts. Cultural differences, for example, have a profound influence on how projects are managed, how decisions are made, and how various stakeholders engage with the process. These cultural factors, together with the influence of international governance frameworks, will have a significant impact on the success of BIM implementation in global projects.

The version of the UBIML-C framework presented in this chapter is an initial conceptual attempt to rethink the adaptation of BIM in the context of global construction projects, especially as these projects move toward more sustainable, digitally integrated practices. It considers the interplay among technology, culture, and governance, and how these factors collectively shape the trajectory of BIM adoption. As the construction industry continues to grapple with the complexities of sustainable development and the global implementation of BIM, this framework offers a foundation for further exploration and refinement.

Researchers and practitioners in the field are invited to build on the concepts introduced in this chapter, expanding upon the UBIML-C framework and testing it in real-world projects. The goal is to encourage the widespread acceptance and application of BIM technologies in sustainable construction projects worldwide, regardless of geographical or cultural barriers. Ultimately, by exploring and understanding the variables that influence BIM adoption in different contexts, the UBIML-C framework aims to support the integration of BIM as a key enabler of sustainability, innovation, and cross-cultural collaboration in the global construction sector. This will, in turn, contribute to the realization of the UN SDGs and the achievement of higher ESG standards in the built environment.

REFERENCES

Alkatheeri, H., Markopoulos, E., & Al-qayed, H. (2023). An organizational and operational capability and maturity assessment for SMEs in emerging markets towards the ESG criteria adaptation. In: T. Ahram & R. Taiar (eds.) *Human interaction & emerging technologies (IHIET 2023): Artificial intelligence & future applications. AHFE (2023) international conference* (AHFE Open Access, Vol. 111). *AHFE International, USA*. http://doi.org/10.54941/ahfe1004075

Deng, M., Menassa, C. C., & Kamat, V. R. (2021a). From BIM to digital twins: A systematic review of the evolution of intelligent building representations in the AEC-FM industry. *Journal of Information Technology in Construction*, 26, 58–83. https://doi.org/10.36680/j.itcon.2021.005

Díaz, J., & Antön, L. Á. (2014). Sustainable construction approach through integration of LCA and BIM tools. *Computing in Civil and Building Engineering* [Preprint]. https://doi.org/10.1061/9780784413616.036

Dulau, E., Botha-Ravyse, C. R., Luimula, M., Markopoulos, P., & Markopoulos, E. (2019). A virtual reality game for cognitive impairment screening in elderly: A user perspective. In: *Proceedings of the 10th IEEE international conference on cognitive InfoCommunications. CogInfoCom 23–25 October 2019. Naples, Italy.* https://doi.org/10.1109/CogInfoCom47531.2019.9089973

Friedman, T. (2007). *The world is flat: The globalized world in the twenty-first century*. Picador.

Ghemawat, P. (2001). Distance still matters. *Harvard Business Review*, 137–147.

Grossi, G., & Trunova, O. (2021). Are UN SDGs useful for capturing multiple values of smart city? *Cities*, 114, 103193.

Grover, P., Kar, A. K., & Ilavarasan, P. V. (2019). Impact of corporate social responsibility on reputation—Insights from tweets on sustainable development goals by CEOs. *International Journal of Information Management*, 48, 39–52.

Hadro, D., Fijałkowska, J., Daszyńska-Żygadło, K., Zumente, I., & Mjakuškina, S. (2022). What do stakeholders in the construction industry look for in non-financial disclosure and what do they get? *Meditari Accountancy Research*, 30(3), 762–785.

Jackson, C. (2021). *What is sustainable construction?* construction21.org. Available at www.construction21.org/articles/h/what-is-sustainable

Jeffrey, S., Rosenberg, S., & McCabe, B. (2019). Corporate social responsibility behaviors and corporate reputation. *Social Responsibility Journal*, 15(3), 395–408.

Jing, W., & Alias, A. H. (2024). Key factors for BIM implementation in the context of ESG and SDGs integration: *A systematic literature*, *16*(21), 9504.

Koller, T., Nuttall, R., & Henisz, W. (2019). Five ways that ESG creates value. *The McKinsey Quarterly*.

LePage, D., & Renaerts, E. (2023). Achieving the UN SDGs through the integration of social procurement in construction projects. *In: The role of design, construction, and real estate in advancing the sustainable development goals* (pp. 129–147). Springer International Publishing.

Liao, L., Zhou, K., Fan, C., & Ma, Y. (2022). Evaluation of complexity issues in building information modeling diffusion research. *Sustainability*, 14, 3005. https://doi.org/10.3390/su14053005

LSEG (2024). *Environmental, Social and Governance scores from LSEG, LSEG Data and Analytics, October 2024*, Available at: www.lseg.com/content/dam/data-analytics/en_us/documents/methodology/lseg-esg-scores-methodology.pdf

Markopoulos, E., Katheeri, H., & Qayed, H. (2023). A decision support system architecture for the development and implementation of ESG strategies at SMEs. In: T. Ahram, W. Karwowski, P. Di Bucchianico, R. Taiar, L. Casarotto, & P. Costa (eds.) *Intelligent human systems integration (IHSI 2023): Integrating people and intelligent systems. AHFE (2023) international conference* (AHFE Open Access, Vol. 69). AHFE International, USA. http://doi.org/10.54941/ahfe1002916

Markopoulos, E., & Luimula, M. (2020). Immersive safe oceans technology: Developing virtual onboard training episodes for maritime safety. *Future Internet*, 12, 80. https://doi.org/10.3390/fi12050080

Markopoulos, E., Markopoulos, P., Nandi, A., Wu, T., Zhao, K., Samkova, M., & Huang, M. (2024). Using the UN SDGs and the ESG Index towards the development of a unified building information modelling language and culture for sustainable construction. *Sustainable Construction in the Era of the Fourth Industrial Revolution*, 149, 1–10.

Markopoulos, E., Markopoulos, P., Nandi, A., Zhao, K., Samkova, M., Wu, T., & Kantola, J. (2024). Aligning digital twins and metaverse with the UN SDGs and applying them to understand human behaviour in smart and virtual cities. In: M. Zallio (eds.) *Accessibility, assistive technology and digital environments. AHFE (2024) international conference*

(AHFE Open Access, Vol. 121). AHFE International, USA. http://doi.org/10.54941/ahfe1004612

Markopoulos, E., Ramirez, A., Markopoulos, P., & Vanharanta, H. (2022). Gamification in a democratic pro-environmental behaviour model towards achieving effective ESG corporate strategies. In: E. Markopoulos, R. S. Goonetilleke, & Y. Luximon (eds.) *Creativity, innovation and entrepreneurship. AHFE (2022) international conference* (AHFE Open Access, Vol. 31, pp. 104–114). *AHFE International, USA.* http://doi.org/10.54941/ahfe1001512

Markopoulos, E., & Ramonda, M. (2022). An ESG-SDGs alignment and execution model based on the Ocean Strategies transition in emerging markets. In: E. Markopoulos, R. S. Goonetilleke, & Y. Luximon (eds.) *Creativity, innovation and entrepreneurship. AHFE (2022) international conference* (AHFE Open Access, Vol. 31, pp. 93–103). *AHFE International, USA.* http://doi.org/10.54941/ahfe1001511

Markopoulos, E., Zhao, K., Samkova, M., & Vanharanta, H. (2023). ESG and UN SDGs driven strategy generation process for green and pink oceans. In: V. Salminen (eds.) *Human factors, business management and society. AHFE (2023) international conference* (AHFE Open Access, Vol. 97). AHFE International, USA. http://doi.org/10.54941/ahfe1003878

Markopoulos, P., & Markopoulos, E. (2023). Addressing the UN 2030 sustainable development agenda and the ESG Index with serious games in virtual environments. In: T. Ahram & C. Falcão (eds.) *Human factors in virtual environments and game design. AHFE (2023) international conference* (AHFE Open Access, Vol. 96). AHFE International, USA. http://doi.org/10.54941/ahfe1003862

Meng, T., Yahya, M. H. D. H., Ashhari, Z. M., & Yu, D. (2023). *ESG performance, investor attention, and company reputation: Threshold model analysis based on panel data from listed companies in China*. Heliyon.

Migilinskasa, D., Popovb, V., Juoceviciusc, V., & Ustinovichiusd, L. (2013). The benefits, obstacles and problems of practical BIM implementation. *Procedia Engineering*, 57, 767–774.

Najjar, M., Figueiredo, K., Palumbo, M., & Haddad, A. (2017). Integration of BIM and LCA: Evaluating the environmental impacts of building materials at an early stage of designing a typical office building. *Journal of Building Engineering*, 14, 115–126.

Opoku, A. (2022). 10. Construction industry and the Sustainable Development Goals (SDGs). *Research Companion to Construction Economics*, 199, 199–214.

Ramnath N. (2011). Pankaj Ghemawat: The world is not flat. *So what. Forbes India. June* 30, 2011. www.forbesindia.com/article/special/pankaj-ghemawat-the-world-is-not-flat-so-what/26292/1

Sovrano, F., & Vitali, F. (2024). Explanatory artificial intelligence (YAI): Human-centered explanations of explainable AI and complex data. *Data Min Knowledge Discovery*, 38, 3141–3168. https://doi.org/10.1007/s10618-022-00872-x

UNDP (2022). *The Government of Borno State and UNDP Reopens the Ngarannam Community in Northeast Nigeria*. United Nations Development Programme. *October* 24, 2022. Available at: www.undp.org/nigeria/press-releases/government-borno-state-and-undp-reopens-ngarannam-community-northeast-nigeria

Visvizi, A., & del Hoyo, R. P. (eds.) (2021). *Smart cities and the UN SDGs*. Elsevier.

Waas, L. (2022). Review of BIM-based software in architectural design: Graphisoft ArchiCAD vs Autodesk Revit. *Journal of Artificial Intelligence in Architecture*, 1(2), 14–22.

10 Beyond Resources

Innovative Use of Mediterranean Materials for Sustainable Design

Raquel Salomé and Ana Cardoso

10.1 INTRODUCTION

The global momentum toward a circular economy diverges from the conventional "take, make, dispose" linear model and has become a critical framework for advancing sustainable and efficient resource practices. Within this paradigm, the Mediterranean region – distinguished by its historical depth, ecological diversity, and wealth of indigenous resources – emerges as an ideal setting for embedding the main principles of circular economy, both in construction techniques and design methodologies. The region boasts an abundance of renewable materials, from cork to ceramics, that are suitable for sustainable applications. However, to realise their full potential we need a more transformative approach to sourcing, processing, and reutilising these materials in ways that are aligned with circular economy goals, emphasising regeneration, waste reduction, and resource efficiency.

Technological advancements, particularly those associated with the Fourth Industrial Revolution, provide essential mechanisms for the innovative renewal of traditional Mediterranean materials. By merging ancestral craftsmanship with contemporary technological methods, local artisans and manufacturers can generate novel, sustainable products that stimulate both environmental preservation and economic development. This dynamic intersection between heritage and innovation opens pathways for creating design solutions that are both ecologically responsible and economically viable, revitalising traditional materials within a modern sustainability framework.

This study presents a project designed to engage collective memory and cultural identity as foundational pillars of design, promoting critical discourse around the role of Mediterranean materials in sustainable construction practices. Central to this project is the development of a modular cladding system for building façades – adaptable to both vertical and horizontal surfaces – that integrates passive and active solar functions while preserving aesthetic cohesion. This design harmonises passive solar features, including solar protection, thermal storage, ventilated walls, and Trombe wall effects, with active technologies such as solar collectors and photovoltaic cells. The resulting solar-responsive cladding system enhances energy efficiency and significantly mitigates a building's energy footprint.

DOI: 10.1201/9781003672036-10

A pivotal aspect of the project is its commitment to using locally sourced Mediterranean materials, exemplified by the incorporation of cork, which enhances sustainability while supporting regional economies. By prioritising indigenous resources, the project exemplifies circular economy principles that prioritise local sourcing, waste minimisation, and resource resilience.

In sum, this study proposes that the integration of circular economy principles with innovative uses of Mediterranean materials and traditional craftsmanship offers a robust model for sustainable development. By rethinking local resources through the lens of cutting-edge recycling technologies, the Mediterranean has the potential to serve as a global exemplar of circular economy success, demonstrating how environmental stewardship can coexist with economic vitality.

10.2 CONTEXTUALIZATION

Since the 1990s, the discourse on sustainability has shifted from a peripheral concern to a more central pillar of the global environmental strategy. This transition is particularly relevant in the architecture, design, and construction sectors, and they account for approximately 40% of global carbon emissions and half of the global energy consumption. Responding to the critical need to reduce carbon emissions by 50%, architects and designers have assumed pivotal roles in promoting sustainable practices. As Bergdoll and Ursprung (2010a) emphasise, a fundamental aspect of this shift is reconceptualising buildings not as static structures, but as integrated systems that can employ passive energy-saving measures alongside active energy-generation technologies. This paradigm combines long-standing bioclimatic principles with more modern technological advancements, achieving both environmental sustainability and an aesthetic value.

A vital yet frequently overlooked aspect of this transformation is the preservation and integration of cultural heritage within sustainable design. The convergence of historical legacies with ecological considerations presents an opportunity to harmonise preservation and innovation. Therefore, increasingly, natural energy sources, particularly solar power, have become central to sustainable architecture. The integration of solar technologies – encompassing both passive and active systems – is marking a critical path toward sustainable building practices (Guzowski and Guzowski 2010). This holistic approach regards the building as an interconnected energy system, leveraging local resources like solar, wind, and geothermal energy to maximise efficiency. According to Guzowski and Guzowski (2010), hybrid solar systems with intelligent, adaptable façades represent a significant step forward, merging cultural and technological elements in order to create buildings that are both efficient and meaningful.

Mediterranean architecture, with its centuries-old synergy between natural elements and design solutions, is prepared to offer a robust foundation for new sustainable practices. Over a millennium, architects and designers in this region have utilised sunlight and wind to inform spatial distribution, material selection, and climate control. Roman advancements, such as south-facing windows designed to retain solar heat, underscore the enduring relevance of natural energy in architectural design for human comfort and environmental performance. Furthermore, geometric considerations in traditional Mediterranean layouts not only provide aesthetic cohesion but also optimise passive solar gain, setting a precedent for more sophisticated architectural responses to environmental conditions.

In the Mediterranean, architecture and design transcend ecological engineering; they are deeply interwoven with cultural and aesthetic identity characteristics. The use of natural resources like sunlight and wind is a practical necessity and an expression of regional heritage. As sustainability becomes a global imperative, architects and designers are revisiting traditional Mediterranean methods to inform contemporary designs that are both energy-efficient and culturally resonant. This renewed focus on solar architecture – incorporating both passive and active elements – demonstrates this belief that a sustainable future is inherently linked to the utilisation of natural, renewable energy sources, such as the sun.

The influence of Arabian architectural traditions also manifests in modern Mediterranean design, particularly in areas like North Africa, southern Spain, and Portugal. For instance, Jean Nouvel's design of the Institut du Monde Arabe in Paris reinterprets traditional Arabian geometric forms in a contemporary setting. The building's dynamic façades feature mobile diaphragms resembling camera shutters, creating shifting patterns of light and shadow that highlight the connection between solar architecture and historical influences. These designs underscore the potential of interactive façades to mediate between interior spaces and the natural environment, illustrating a synthesis of cultural heritage with advanced solar technologies. Another example is Mario Botta's design of the San Francisco Museum of Modern Art, which reflects Mediterranean architectural traditions with its use of simple geometric forms, heavy stone construction, and a focus on light and shadow, evoking the timeless qualities of Mediterranean building techniques.

Today, solar architecture transcends the mere installation of solar collectors; it should embody a holistic design philosophy that integrates local resources with site-specific factors such as climate, orientation, spatial organisation, and material selection. Among these materials, ceramics have gained prominence due to their environmental benefits, versatility, and aesthetic appeal. Traditionally used in Mediterranean architecture for structural and decorative purposes, ceramics now serve as essential materials in sustainable construction, allowing architects and designers to develop innovative and customised projects.

The exploration of modular design methodologies has furthered the understanding of sustainable architecture. A modular system may offer more flexibility and adaptability, enabling buildings to respond dynamically to environmental changes and user needs. In the realm of solar architecture, modular design has inspired multifunctional cladding systems, such as ceramic-based products that incorporate both passive and active solar features. These systems combine traditional Mediterranean geometries with advanced technologies for solar protection, thermal storage, and ventilation, along with active photovoltaic panels. These multifunctional façades not only improve energy efficiency but also can foster biodiversity by creating habitats for local plants and birds – essentially contributing to the regeneration of the ecosystem.

The integration of Mediterranean cultural elements into contemporary design is exemplified by such projects that are able to merge solar technologies with modular systems. Such designs provide advanced solutions for building façades that are both practical and rooted in local heritage. A prototype of this modular system has been created and tested, demonstrating its potential as a sustainable solution bridging traditional craftsmanship and modern engineering. By merging design, technology, and cultural history, this project illustrates the transformative potential of solar architecture within the Mediterranean context.

The future of solar architecture may very well lie in the integration of intelligent, self-regulating systems that respond to environmental conditions in real time. Hegger (2003b) observes that the next wave of architectural innovation will focus on hybrid systems that incorporate passive solar strategies alongside active technologies, such as climate-responsive buildings and auto-regulating façades. This approach promises to revolutionise architectural design, creating structures that are not only energy-efficient but also capable of interacting dynamically with their surrounding environment.

The culmination of this project results from an intricate and rigorous design process, marked by comprehensive stages of surveying, analysis, and the development of communication proposals, all grounded in a deep commitment to sustainable design principles. Beyond mere formal manipulation, the project seeks to achieve a conceptual foundation that integrates seamlessly with various disciplines, particularly design, architecture, and engineering, with a strong emphasis on sustainability characteristics. Positioned within its operational context, the designer may engage not only with the rich cultural heritage embedded in centuries of material culture but also with the evolving sociocultural models and emerging social dynamics, all while maintaining a focus on new sustainable trajectories.

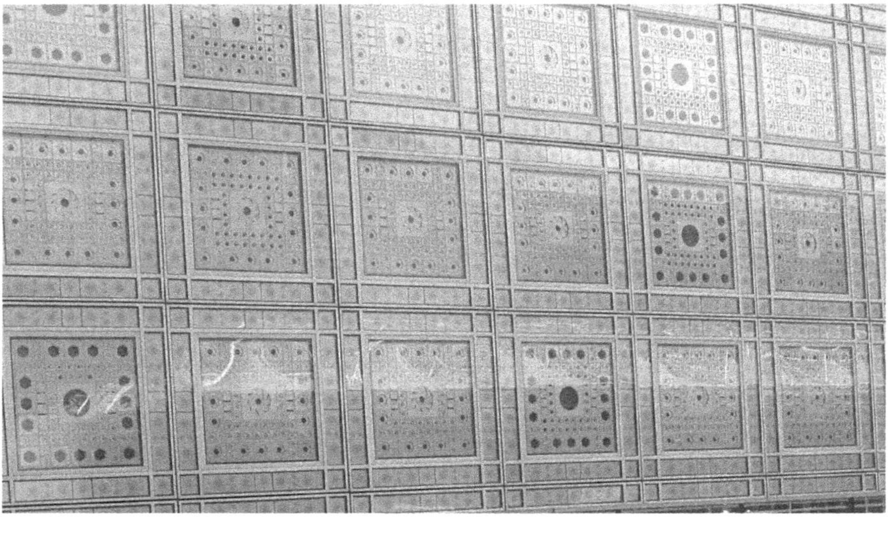

FIGURE 10.1 The Institut du Monde Arabe, by Jean Nouvel, Paris, France, 1987. Photograph by the author.

This project adopts a distinctive condition, allowing its intervention to influence societal structures and values, guided by sustainability as a core principle. It possesses the potential to transform functional uses, alter established practices, evoke emotional responses, and reshape relationships between individuals and their built environment. In this transformative process, the designer should act as a mediator within the dynamics of societal groups, imparting the project with a unique language, worldview, emotional undertones, cultural values, and individual identity, all shaped by sustainability needs.

A comprehensive analysis of Mediterranean architecture, with a specific focus on sustainability, easily highlights the crucial role of geometry in both structural and ornamental components. Of particular interest is the use of hexagonal meshes, as explored by Castéra and Peuriot (1996). The hexagon, akin to the square and triangle, is a highly efficient geometric unit, capable of filling a plane without gaps, thus promoting both efficiency and minimising environmental impact.

Islamic patterns, as demonstrated in historical wood carvings and paintings (see Figure 10.2a, b), may exemplify the integration of cultural and geometric elements within sustainable design practices. These patterns, such as the 12-pointed star inscribed within a hexagonal grid, were very much instrumental in the ideation process, guiding the development and conceptualisation of the final geometry that supports the integration of energy-efficient technologies within the modular product system. While the project predominantly focuses on hexagonal geometries, it remains open to exploring non-hexagonal forms intrinsic to Arabic architectural traditions through a sustainability-oriented lens. One such alternative examined is the "mudejar" module, known for its characteristic geometric motifs, often referred to as "cock's feet" (Meco, 1993).

Building upon this geometric and sustainable foundation, three primary modules were conceptualised: "*alfardon*," "*alicatado*," and "*mudejar*" (Figure 10.3). The intricate integration of cultural and sustainable design references enriches the project aesthetically while reaffirming its commitment to preserving and reinterpreting historical architectural elements within the framework of contemporary sustainable design practices.

One notable reference is the "*alfardon*," a hexagonal pavement tile that historically was used in Valencia, Spain, and Italy during the 15th and 16th centuries. Prominent in mosaic pavements by the mid-15th century, particularly in the palaces of Lisbon and Beja (south of Portugal), these tiles – shaped by Islamic design principles – were arranged in intricate geometric patterns. Within the scope of this project, the aim was not to introduce a new geometry pattern but to distil the essence of Islamic art's geometric abstraction and repurpose it. The focus was on interpreting, recreating, and re-contextualising these forms within contemporary settings, highlighting the advantages of the selected geometry for material applications, fixation systems, and the integration of various eco-energy and sustainable functions during the conceptualisation phase.

However, challenges emerged regarding the performance of the chosen geometry, specifically in relation to the thermodynamic reliability sought in the project's objectives. To address this concern, a real prototype was experimentally tested, and

FIGURE 10.2 (a) Door of the Palais Royal, Casablanca, Morocco; (b) Identification of secret geometries of the Bahía Palace, Marrakech, Morocco. (Created by the author.)

its performance was compared to that of a conventional commercial system. The findings from this experimental evaluation are discussed within this paper.

The technical innovation resulting from this project is the Eco-Multitask Modular Alveoli (EMMA), a hexagonal-based modular system designed with integrated energy and aesthetic functions. The EMMA tile represents a harmonious fusion

FIGURE 10.3 The "*alfardon*", "*alicatado*" and "*mudejar*" modules, 2012. (Rendered by Rafael Coelho.)

of design and technology, offering a versatile, eco-efficient solution for building cladding. This hybrid technological system incorporates a range of eco-efficient functionalities, with a particular emphasis on thermal performance, which serves as the core component. Furthermore, the EMMA tile integrates additional functionalities, including photovoltaic, biological, and lighting elements (Figure 10.4a). The ventilated function, driven by the chimney effect, is facilitated by a fully enclosed ceramic module. Strategically placed vents on the thermodynamic module allow air to circulate, enabling the chimney effect, where warmer air rises from the top, creating a flow that moves cooler air from the bottom.

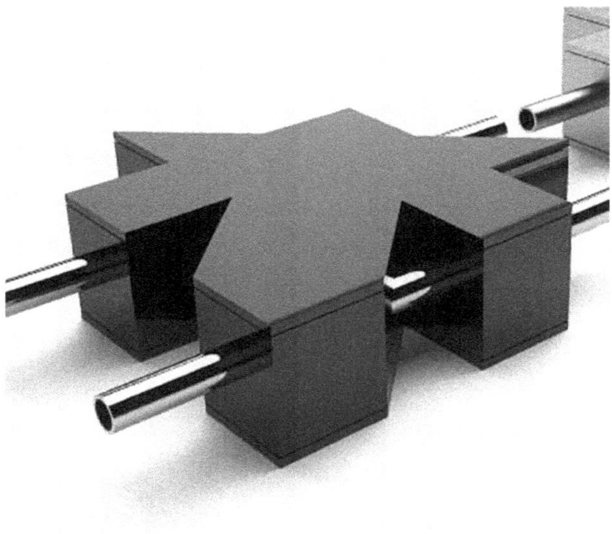

FIGURE 10.3 (Continued)

This modular assembly system consists of distinct technical alveoli, designed for high performance. The system is adaptable for both new construction projects and the rehabilitation of existing structures, reinforcing its sustainable credentials. By integrating energy-efficient technology with a modular design inspired by cultural precedents, the EMMA tile exemplifies a forward-thinking approach to sustainable architecture, balancing innovation with a respect for historical and cultural legacies.

This project proposes five distinct types of modules, each designed with specific functions that address sustainability through the integration of advanced technologies and traditional materials. The modules are as follows:

1. Passive thermal function: This module includes a ceramic component designed to utilise the chimney effect, as well as a glass module that operates on the principle of the wall effect (*trombe* walls), which facilitates passive heating and cooling strategies.
2. Active thermal function: This module is configured as a solar collector, designed to produce domestic hot water (DHW) through the conversion of solar energy, contributing to the energy efficiency of the building.
3. Lighting functions: An open module that filters natural light to create a balance between natural illumination and energy conservation, enhancing indoor environmental quality.

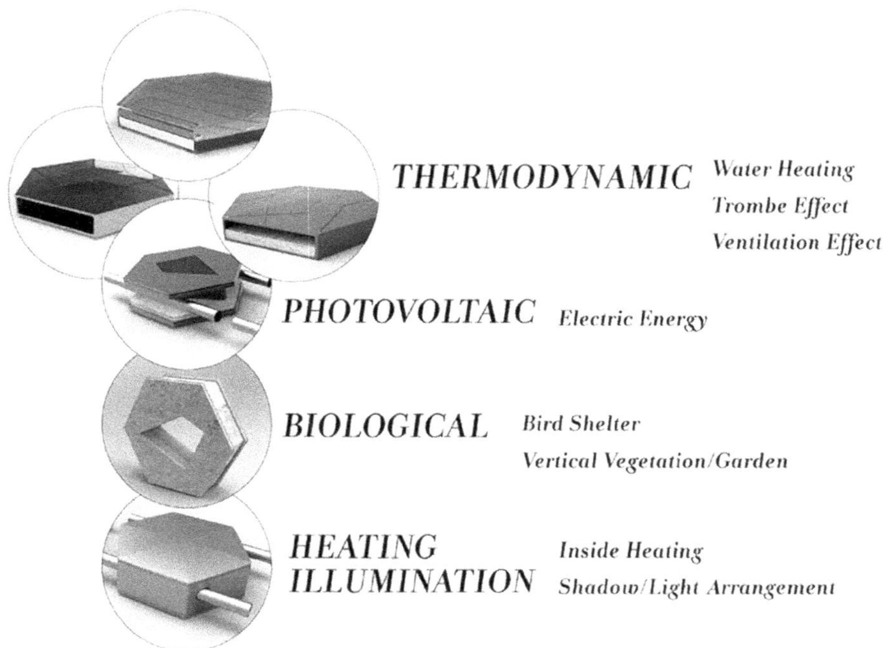

FIGURE 10.4a Function study of the EMMA system, 2012. (Created by author and Rendered by Rafael Coelho.)

4. Biological functions: This open module is also designed to filter natural light, while simultaneously creating spaces that support biological processes, such as green gardens or habitats for small bird species (Figure 10.4b).
5. Electrical function: A photovoltaic module designed to convert solar energy into electrical power, aligning with the overarching sustainability goals of the project.

The development of these modules is illustrated in Figure 10.4a, which demonstrates the interconnectedness of materials such as ceramics, glass, copper, and cork. These materials are integrated into both thermodynamic and photovoltaic systems, underscoring the project's commitment to sustainability. The modular assembly of these components, likened to the structure of "lego"-like parts, allows each piece to incorporate specific technological and aesthetic functions. This method emphasises eco-efficiency, promoting environmentally responsible design that is aligned with the principles of sustainable development.

A consistent geometric framework is employed throughout the design, where the upper surfaces of the modules can be replaced with glass to facilitate the passive thermal *Trombe* function, which optimises heat storage and release. Multiple module variations were developed, each incorporating distinct energy systems to support both

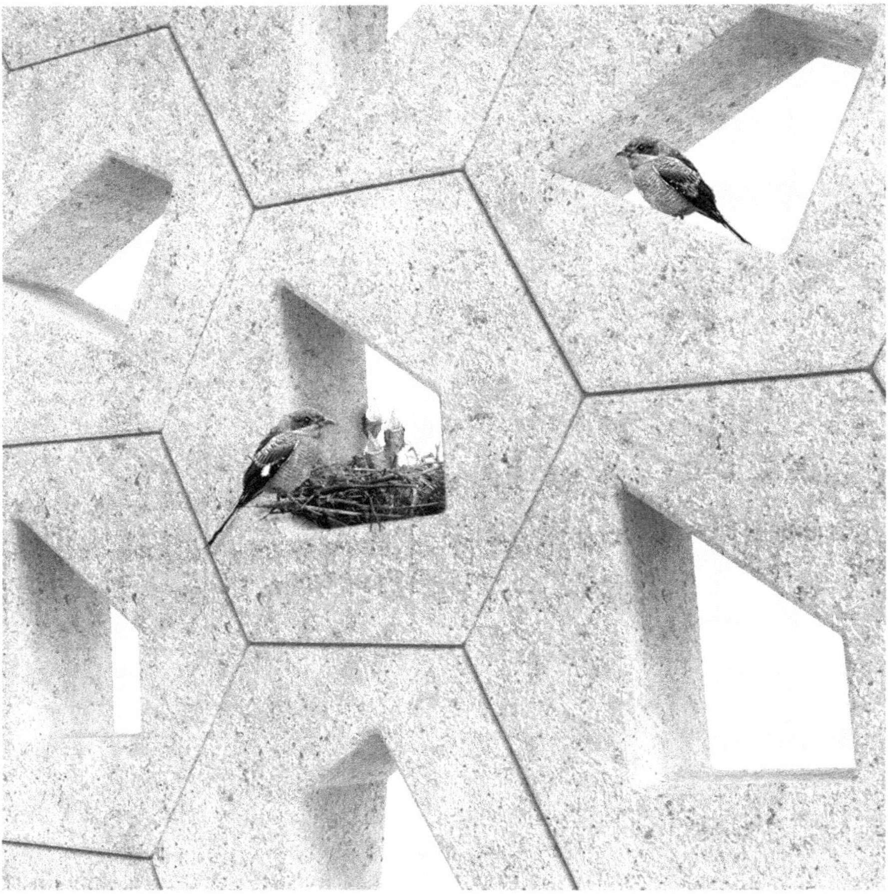

FIGURE 10.4b Biological functions module – designed to filter natural light or support biological processes, such as green gardens or natural habitats for small bird species, 2012. (Rendered by Rafael Coelho and Elias Marques.)

thermodynamic and photovoltaic capabilities. In particular, the biological and illumination modules are characterised by central apertures: the illumination module is specifically designed to filter natural light, enhancing daylighting strategies within buildings, while the biological module is intended to create mini-habitats conducive to the growth of plants or the habitation of small avian species, contributing to urban biodiversity.

The schematic overview in Figure 10.5 further details the principal modules developed within this framework. Each of these modules contributes to a holistic approach to sustainable architecture, integrating technological advancements with natural processes and environmental responsiveness. Through the integration of these systems, the project demonstrates how a multifunctional and modular design

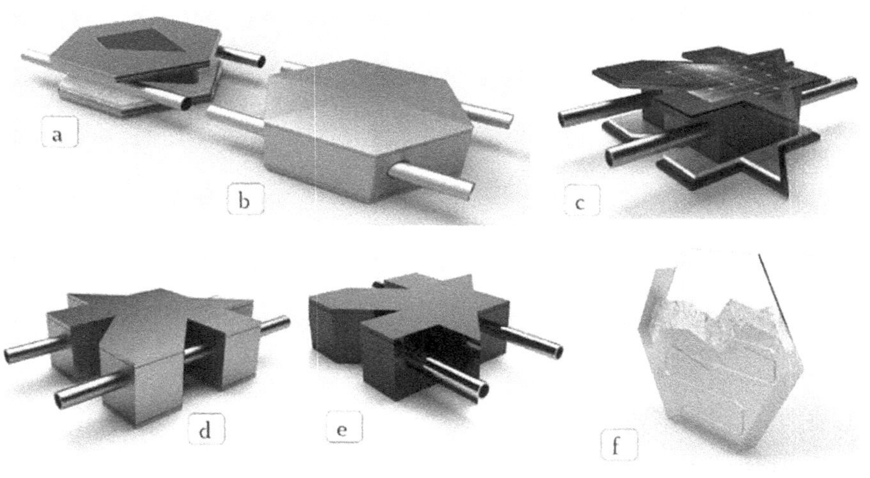

FIGURE 10.5 Technical modules: (a) opened "*alfardon*" tile made of black ceramic; (b) copper "*alfardon*" tile; (c) photovoltaic "*mudejar*" tile; (d) copper "*mudejar*" tile; (e) black ceramic "mudejar" tile; and (f) thermodynamic "*alfardon*" tile, 2012. (Rendered by Rafael Coelho.)

can achieve energy efficiency, environmental resilience, and alignment with contemporary sustainable design practices.

10.3 THE ECO-MULTITASK MODULAR ALVEOLI (EMMA) AND THE DIFFERENT TECHNICAL OPTIONS

The Eco-Multitask Modular Alveoli (EMMA) project presents a design concept that merges innovative design with technological advancements, creating a multifunctional, eco-efficient system for cladding buildings. EMMA is conceptualised as a hybrid technological system, integrating various sustainable functions, with the primary emphasis placed on thermal performance. In addition to thermal management, the system also incorporates photovoltaic, biological, and illumination functions, as illustrated in Figure 10.6.

The EMMA system is built around a modular design consisting of various technical alveoli optimised for high performance. These modules are versatile and applicable in both new constructions and the renovation of existing buildings, providing a flexible solution to enhance overall building performance. At the heart of the system's technological innovation is its hexagonal-based geometry, which serves the dual purposes of aesthetics and functionality. A key feature of the system is its ability to facilitate the ventilated function, or chimney effect, through a fully enclosed ceramic module. Vents are strategically placed at the top, where temperatures are elevated, and at the bottom, where temperatures are lower, creating natural airflow. These vents

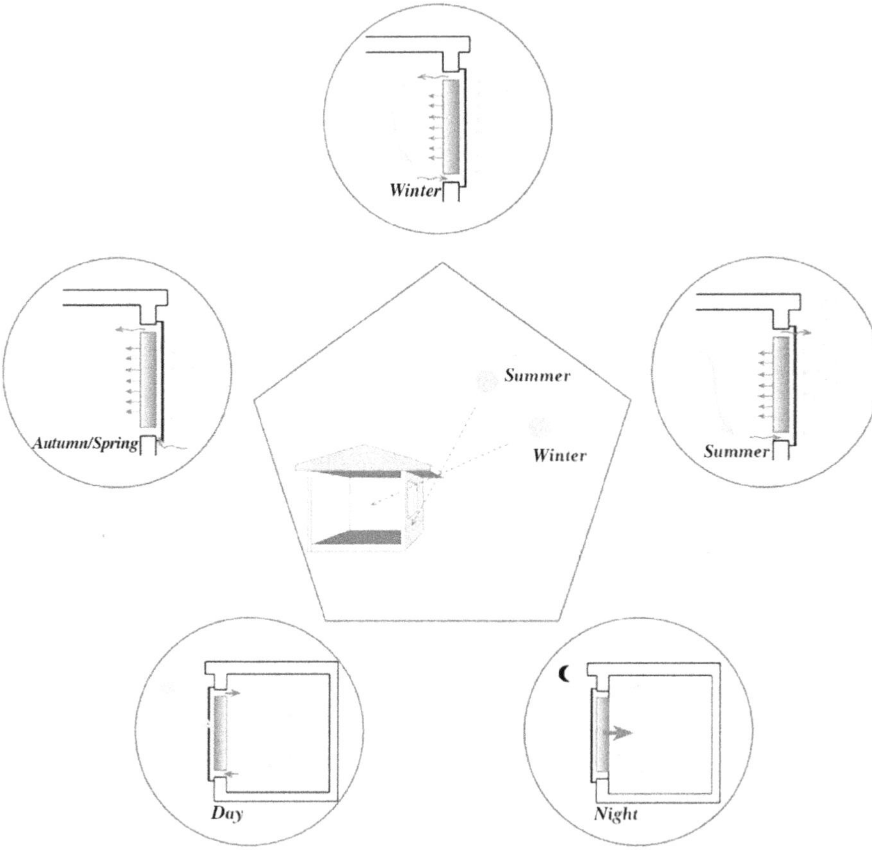

FIGURE 10.6 EMMA Functions, 2012. (Created by the author.)

can be controlled either by thermostatically regulated motors or manually adjusted, offering flexibility in thermal management.

In an alternative design, the upper surface is replaced with glass to enable the passive thermal *Trombe* effect, optimising heat retention and release within the building. Building on this concept, additional modules have been developed, each incorporating distinct energy systems that support both thermodynamic and photovoltaic functionalities, further enhancing the system's energy efficiency.

The design also includes biological and illumination modules, which feature open centres to filter natural light and foster biodiversity. The biological module is capable of supporting green gardens and providing environments for small bird species, promoting urban biodiversity and environmental quality. Meanwhile, the illumination module ensures the efficient use of natural light, reducing energy consumption related to artificial lighting (Figures 10.7 and 10.8). These modules can be arranged in various configurations based on the building's façade orientation, optimising both thermal performance and energy efficiency.

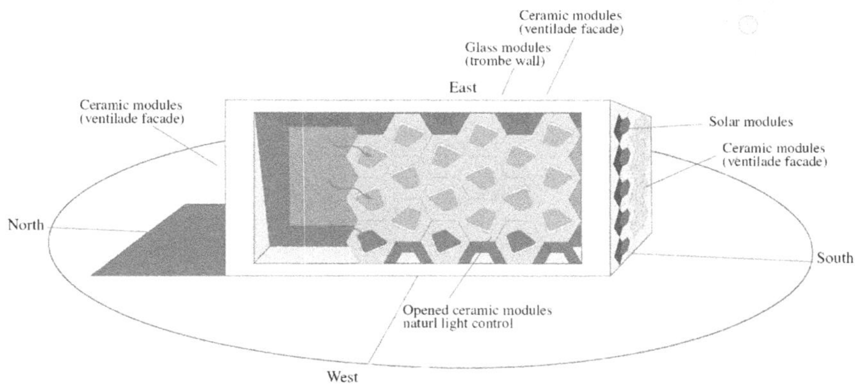

FIGURE 10.7 Building façade orientation with modules assembled in different, 2012. (Created by the author.)

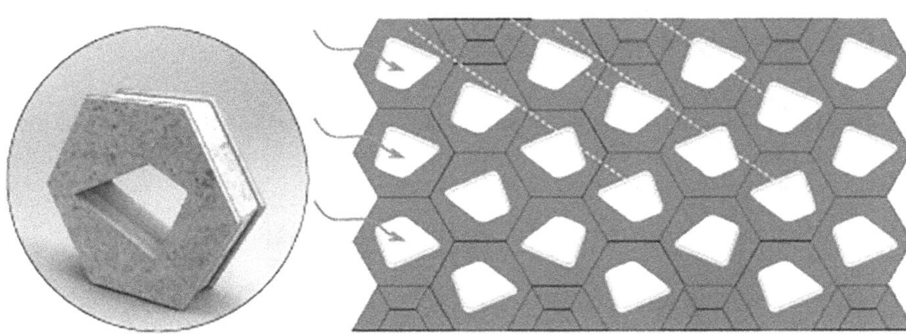

FIGURE 10.8 Flexible solar protection to allow more solar penetration in winter, 2012. (Created by author and Rendered by Rafael Coelho.)

As shown in Figure 10.9, the glass-based thermal module can perform both *Trombe* and chimney functions, offering versatile solutions for passive heating and cooling. This adaptability allows the system to be customised for different climates and building orientations, ensuring optimal performance throughout the year.

Figure 10.10 highlights how different materials, including ceramics, glass, copper, and cork, were integrated into the alveoli's design. These materials are selected for their compatibility with both thermodynamic and photovoltaic solar functions. The modular assembly is akin to "lego" blocks, with each part engineered to fulfil specific technological and aesthetic requirements. This method underscores the system's

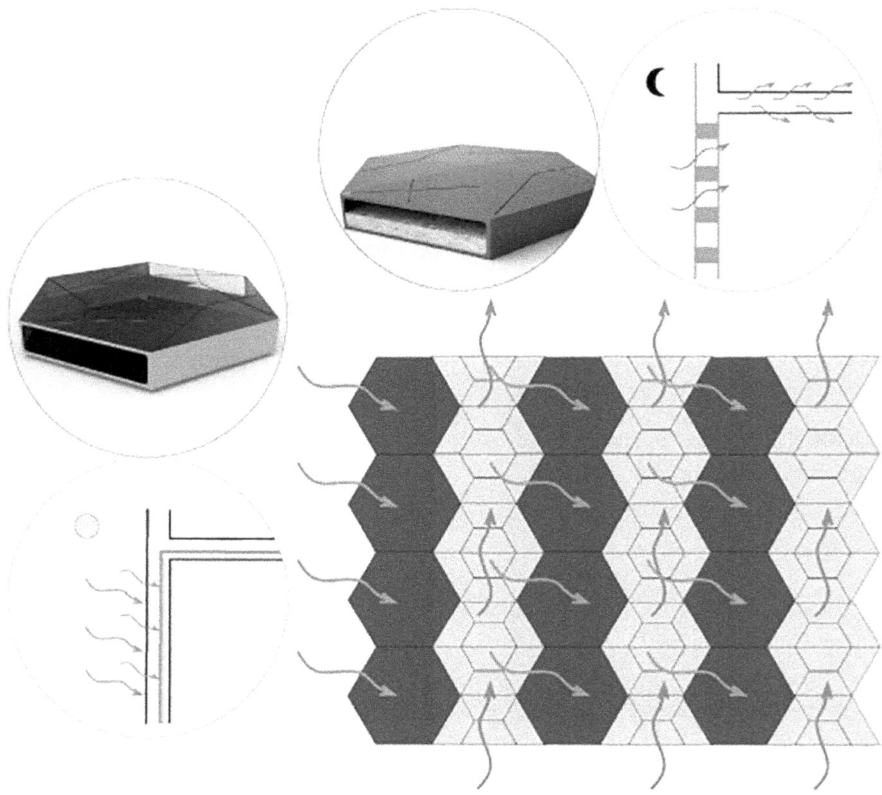

FIGURE 10.9 Thermal passive glass module to perform Trombe and chimney functions, 2012. (Created by author and Rendered by Rafael Coelho.)

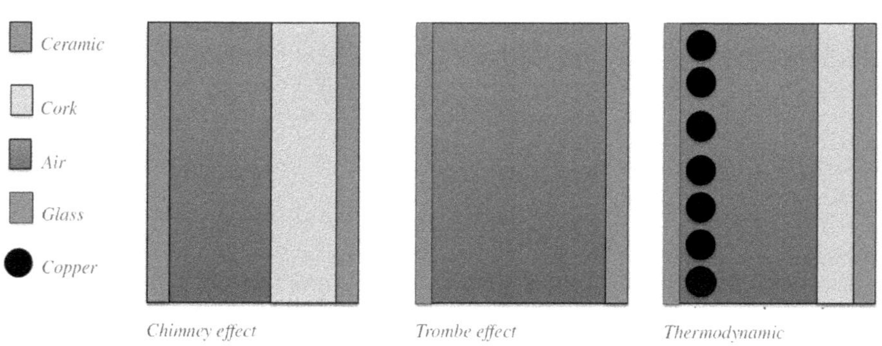

FIGURE 10.10 Organisation of the materials for different energy functions, 2012. (Created by the author.)

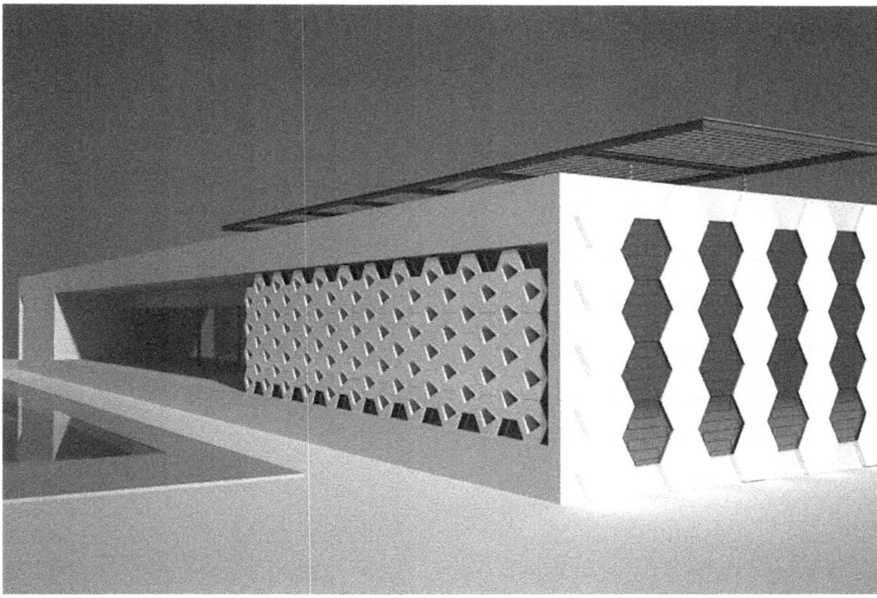

FIGURE 10.11 External view of a façade composed with ceramic and solar thermal modular and illumination parts, 2012. (Rendered by Nuno Martins.)

inherent flexibility and its ability to tailor eco-efficient functionalities to meet various building needs.

The EMMA system enables the construction of thermodynamic, self-sustaining façades composed of modules with diverse functions, which can be arranged to suit the specific demands of the building's design and environmental conditions. Figures 10.11 and 10.12 offer external and internal views of a building façade composed of ceramic and solar thermal modules, showcasing how the system integrates both energy and design considerations.

In conclusion, the EMMA project exemplifies the fusion of sustainable design principles with advanced technological innovations. By integrating materials such as ceramics, glass, and copper with functions like solar energy capture, thermal regulation, and biodiversity enhancement, the system provides a comprehensive solution for eco-efficient building design. Its modular, customisable nature ensures that it can be adapted to a variety of architectural contexts, making it a valuable contribution to contemporary sustainable architecture practices.

10.4 MATERIALS AND TECHNOLOGICAL FUNCTIONS

Today the planning and construction of buildings needs to prioritise the use of energy efficiency systems from the outset, with material selection playing a decisive role in shaping the overall energy performance of the building. A building designed with energy efficiency in mind allows for the optimal use of energy resources, particularly

FIGURE 10.12 Internal view of a façade composed with modular parts for inside illumination, 2012. (Rendered by Nuno Martins.)

in relation to heating and cooling systems, which in turn enhances occupant comfort while substantially reducing energy consumption and the associated costs of gas or electricity. In this context, the materials and technologies utilised in the construction process are fundamental to the building's thermal performance and energy efficiency.

Ceramic materials, renowned for their durability and thermal properties, have become integral to modern cladding systems, especially in the design of ventilated façades. These materials are increasingly preferred by architects, designers, and engineers due to their ability to significantly enhance the thermal performance of buildings. The use of ceramic in ventilated façades gained prominence in temperate climates, particularly in Germany, following the energy crises of the 1970s, when the primary objective was to improve building envelopes by enhancing thermal insulation and reducing heat loss. Ventilated façades are distinguished by the presence of an air chamber that facilitates air circulation through the building envelope, driven by temperature gradients and wind action.

A ventilated façade system can incorporate a range of cladding materials, such as large ceramic tiles, along with natural stones such as marble and granite. Large ceramic tiles, in particular, have become a staple in contemporary European construction due to their aesthetic potential and functional advantages for façade design, as discussed by Guzowski (2000). These tiles are typically used to anchor cladding materials to external walls, forming a multi-layered system that provides effective thermal insulation. Ventilated façades offer notable advantages in energy performance, particularly in their ability to regulate heat transfer between the external environment and the interior spaces.

During the summer, the external layer of the ventilated façade acts as a barrier to prevent excessive heat penetration. The residual heat that enters the air chamber raises the temperature of the air, causing it to rise due to the chimney effect. Behind this chamber, insulation material is installed against the wall, creating an additional thermal barrier that further reduces heat transfer. This multi-layered system is highly effective in mitigating heat infiltration, thus enhancing energy efficiency. In contrast, during the winter, the ventilated façade system functions to retain internal heat. The movement of air within the chamber helps to maintain the insulation's dryness and effectiveness, preventing heat loss from the interior of the building (Guzowski, 2000).

Research on glazed solar chimneys for heat recovery in naturally ventilated buildings has demonstrated that maximising ventilation rates in cold winters requires the use of double or triple glazing. However, the inclusion of heat pipes in solar chimneys for heat recovery can increase flow resistance and reduce thermal buoyancy, thereby diminishing overall system efficiency (Liu & Feng, 2012). Within the context of ventilated façades, the temperature differential between the interior and exterior of the building facilitates air movement, which in turn helps to prevent condensation by expelling excess moisture from the insulation material. Furthermore, the continuous application of insulation is critical to avoid thermal bridging, particularly at building corners and junctions. The air chamber in ventilated façades also serves as a barrier to water penetration, further enhancing the durability and efficiency of the building envelope.

Solar energy harvesting plays a pivotal role in enhancing the energy efficiency of buildings, particularly for heating purposes. Liu and Feng (2012) note that passive solar techniques, including Trombe walls, solar chimneys, and solar roofs, have the potential to reduce annual heating demand by up to 25%. Trombe walls, in particular, have been shown to decrease a building's energy consumption by as much as 30% (Hordeski, 2011). These walls function as thermal storage systems, capturing solar energy during the day and gradually releasing it to aid in the ventilation, heating, and cooling of buildings. The environmental benefits of such systems are considerable, contributing to the reduction of greenhouse gas emissions and the mitigation of broader environmental degradation.

In conclusion, the strategic selection of materials, coupled with the integration of advanced technological functions such as ventilated façades and passive solar devices, is integral to achieving energy efficiency in buildings. By utilising natural materials, such as ceramics, in conjunction with innovative technologies such as Trombe walls and solar chimneys, it is possible to significantly reduce energy consumption, promote environmental sustainability, and enhance the comfort and well-being of building occupants.

10.5 THE BETA PROTOTYPE FOR THE *ALFARDON* MODEL

The prototype centres on the development and evaluation of a beta thermodynamic prototype for a solar thermal module (STM) designed to optimise hot water production. The primary objective of this investigation is to assess the feasibility of integrating novel materials and geometric configurations into solar thermal systems, offering potential alternatives to conventional systems without compromising efficiency. This research is situated within the broader context of advancing solar energy applications through innovative design while ensuring functional equivalence with existing commercially available systems.

The prototype fabrication was carried out by FirstStep, a company based in Aveiro, Portugal, with meticulous attention to dimensional design. Each hexagonal module has a surface area of 0.5 m^2 and a side length of 440 mm, enclosed within a glass structure secured to a metallic frame, with a panel thickness of 85 mm. Design calculations were carefully conducted to ensure that the total sunlight capture area matched that of traditional solar thermal systems, thereby facilitating a reliable and consistent comparative analysis of performance metrics.

Experimental trials of the STM prototype were conducted over a one-month period, from 15 April to 15 May 2011, during which comprehensive data on the thermal performance of a 150 litre water tank was collected. The prototype, with a total surface area of 1 m^2, was positioned vertically at a 90 degree angle to the horizontal plane, with the absorbing surface oriented toward the south (azimuth 180 degrees). The collected data focused on daily water temperature fluctuations, allowing for the computation of energy transfer from the solar module to the tank. By calculating the temperature differentials and the fluid accumulation within the system, the experiment approximated the total energy transferred during the trial period.

Figure 10.13 illustrates the thermodynamic heating water beta prototype, highlighting its structural features and alignment. This visual representation of the experimental setup provides key insights into the design considerations that drove the prototype's development. Furthermore, Figure 10.14 presents a comparative graphical analysis juxtaposing simulated and experimental data. This analysis serves to validate the accuracy of the energy production predictions made for the one-month trial, substantiating the efficiency of the prototype's novel design. The findings demonstrate that the introduction of new materials and configurations did not lead to a marked decrease in the system's performance relative to conventional solar thermal technologies.

This research presents a comprehensive case study on the functional equivalence of innovative solar module designs. It highlights the potential for optimising solar energy applications through material and geometric modifications, paving the way for further research and development in the field of renewable energy systems.

10.6 CONCLUSION

In conclusion, this study tries to highlight the transformative potential that integrating circular economy principles with the innovative application of Mediterranean materials can offer, especially as a pathway to a more sustainable, efficient, and ecologically cohesive future in design and architecture. Through a deep appreciation for traditional craftsmanship and a thoughtful reinterpretation of culturally rooted architectural symbols, this research demonstrates how the Mediterranean region can

FIGURE 10.13 "*Alfardon*" thermodynamic heating water beta prototype, 2012. Photograph by the author. (Created by FirstStep.)

serve as a successful model for current circular economy practices. The strategic integration of diverse, locally sourced materials with advanced recycling technologies underscores the region's potential to harmonise environmental responsibility with local economic growth capacities.

The core objective of this research has been to promote the revitalisation and reinterpretation of symbolic Mediterranean forms, materials, and concepts that may very well define the future of the region's cultural identity. This exploration led to the development of an innovative modular building tile system that can be specifically tailored for contemporary architectural contexts. Designed with a strong emphasis

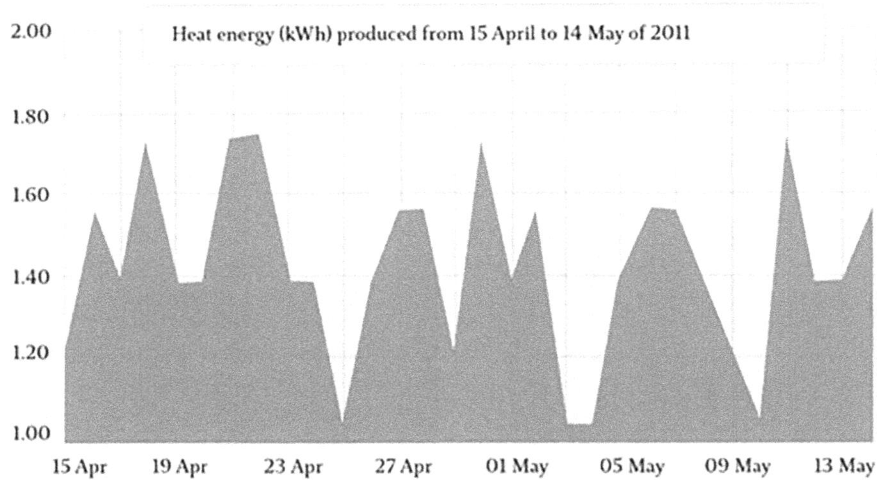

FIGURE 10.14 Graphic showing experimental data measured energy supplied to the 150 litre vessel in the period of time considered, 2012. (Created by FirstStep.)

on sustainability, these tiles embody the principles of natural, renewable, and non-polluting material use, thereby advancing the broader aims of circular and sustainable architecture and design.

To operationalise these concepts, a thermodynamic prototype was developed and subjected to testing. Experimental results indicate that this prototype performs well compared to conventional systems, confirming its efficiency and reliability. Its application potential, especially on vertical façades in winter, provides additional advantages by optimising the façade's interaction with the solar angle, thereby enhancing thermodynamic efficiency.

Ultimately, this research aims to provide a framework for the convergence of sustainable design, cultural heritage, and circular economy principles within modern architectural practice. By anchoring innovation in the Mediterranean's rich cultural legacy, it offers a compelling model for future projects that aim to unify environmental and cultural sustainability, advancing global discourse on sustainable architecture.

ACKNOWLEDGEMENTS

The authors would like to acknowledge the fundamental contributes of António Torres Marques, Faculty of Engineering, University of Porto, Portugal; José António Simões, ESAD – College of Art and Design, Portugal; and Cláudio Torres, Campo Arqueológico de Mértola, Portugal. As for images of 3D modelling: Rafael Coelho and Nuno Martins. Our sincere appreciation to all the invaluable contributions given.

And also the following institutions: ESAD – College of Art and Design, Portugal; ESAD-idea – Research in Design and Art, Portugal; FCT – Fundação para a Ciência e a Tecnologia, Portugal; and BDA – Blue Design Alliance, Portugal.

BIBLIOGRAPHY

Albanese, M. V., B. S. Robinson, E. G. Brehob, and M. K. Sharp. "Simulated and Experimental Performance of a Heat Pipe Assisted Solar Wall." *Solar Energy* 86 (2012): 1552–1562.

Alomar, M. A. "*Creativity in Architecture and Management*". *6th Asian Design International Conference*, Tsukuba, Japan, October 14–17, 2003.

Asnaghi, A., and S. M. Ladjevardi. "Solar Chimney Power Plant Performance in Iran." *Renewable and Sustainable Energy Reviews* 16 (2012): 3383–3390.

Bergdoll, B., and P. Ursprung. *Sauerbruch Hutton*, Vol. 52. Barcelona: Gustavo Gili (2G), 2010a.

Bergdoll, B., and P. Ursprung. *Architecture and Climate: A History of Environmental Design*. New York: Princeton Architectural Press, 2010b.

Broadbent, G. *Design in Architecture*. New York: John Wiley & Sons, 1975.

Castéra, J. M. *Arabesques: Art Décoratif au Maroc*. Paris: ACR Édition, 1996.

Castéra, J. M., and C. Peuriot. *Arabesques: Decorative Art in Morocco*. Paris: Éditions du Chêne, 1996.

Chan, H. Y., S. B. Riffat, and J. Zhu. "Review of Passive Solar Heating and Cooling Technologies." *Renewable and Sustainable Energy Reviews* 14 (2010): 781–789.

El-Said, I., and E. Parman. *Geometric Concepts in Islamic Art*. London: Islam Festival Publishing Company Ltd, 1976.

Fang, X., and T. Yang. "Regression Methodology for Sensitivity Analysis of Solar Heating Walls." *Applied Thermal Engineering* 28 (2008): 2289–2294.

Fiaschi, D., and A. Bertolli. "Design and Energy Analysis of Solar Roofs: A Viable Solution with Esthetic Appeal to Collect Solar Heat." *Renewable Energy* 46 (2012): 56–61.

Gan, G., and R. B. Riffat. "A Numerical Study of Solar Chimney for Natural Ventilation of Buildings with Heat Recovery." *Applied Thermal Engineering* 18 (1998): 1171–1187.

Gershenson, J. K., G. J. Prasad, and Y. Zhang. "Product Modularity: Definitions and Benefits." *Journal of Engineering Design* 14 (2003): 295–313.

Guzowski, M. *Towards Zero-Energy Architecture: New Solar Design*. London: Laurence King, 2010.

Guzowski, M., and M. Guzowski. *Towards a Solar Architecture: Ecological Design for a Sustainable Future*. New York: Routledge, 2010.

Guzowski, S. *Daylighting for Sustainable Design*. New York: McGraw-Hill, 2000.

Hami, K., B. Draoui, and O. Hami. "The Thermal Performances of a Solar Wall." *Energy* 39 (2012): 11–16.

Hegger, M. "From Passive Utilization to Smart Solar Architecture." In *Solar Architecture*, edited by Christian Schittich. Basel: Birkhäuser, 2003a.

Hegger, M. *Intelligent Solar Architecture: Hybrid Systems and Microclimate-Responsive Designs*. Basel: Birkhäuser, 2003b.

Holloway, D. R. *Sun Tempered Architecture A Simple Design Methodology for Passive Solar Houses*. Greenspur, 2011. Available at: ww.dennisrhollowayarchitect.com/SimpleDesignMethodology.html.

Hordeski, M. F. *New Technologies for Energy Efficiency*. New York: The Fairmont Press, 2003.

Hutton, S. *2G N.52*. Barcelona: Editorial Gustavo, 2009.

Jablan, S. V. "*Modularity in Art*". *Visual Mathematics Journal*, 1997. Available at: www.mi.sanu.ac.rs/vismath/jablan/d3.htm

Koyunbaba, B. K., and Z. Yilmaz. "The Comparison of Trombe Wall Systems with Single Glass, Double Glass, and PV Panels." *Renewable Energy* 45 (2012): 111–118.

Kreith, F. "Heat and Mass Transfer". In *Mechanical Engineering Handbook*, edited by Frank Kreith. Boca Raton: CRC Press LLC, 1999.

Liu, Y. W., and W. Feng. "Integrating Passive Cooling and Solar Techniques into Existing Buildings in South China." *Advanced Materials Research* 368–373 (2012): 3717–3720.

Meco, J. *O Azulejo em Portugal*. Lisbon: Edições Alfa, 1989.

Meco, J. *Mudejar Ornamental Tiles from Portugal*. 1993.

Minguet, J. M. *Façades*. Barcelona: Ediciones Monsa, 2010.

Mira, J. *Moldear Ensamblar Proyectar. La Cerámica en Arquitectura*. Spain: ASCER, 2006.

Oliveira, R. S., Coelho, R., A. T. Marques, C. Torres, and J. A. Simoes. "Design of Active and Passive Solar Elements for Sustainable Contemporary Architecture." *Journal of Architectural Engineering Technology* 3 (2014): 132. https://doi.org/10.4172/2168-9717.1000132.

Oliveira, R. S., M. Condesso, C. Farinhas, A. T. Marques, J. Simões, and C. Torres. "Conceção de Módulos Energéticos para Revestimento de Edifícios (I Parte)." *TecnoMetal*, no. 198 (January/February 2012a): 12–20.

Oliveira, R. S., M. Condesso, C. Farinhas, A. T. Marques, J. Simões, and C. Torres. "Conceção de Módulos Energéticos para Revestimento de Edifícios (II Parte)." *TecnoMetal*, no. 199 (March/April 2012b): 8–14.

Orr, D. *The Nature of Design: Ecology, Culture and Human Intention*. Oxford: Oxford University Press, 2002.

Pereira, A., and K. Tschimmel. "The Design of Narrative Jewelry as a Perception-on-Action Process." In *Design Creativity 2012: Proceedings of the 2nd International Conference on Design Creativity*, 2012, 45–56. Kyoto, Japan: The Japan Society of Mechanical Engineers.

Quesada, G., D. Rousse, Y. Dutil, M. Badache, and S. Halle. "A Comprehensive Review of Solar Facades: Opaque Solar Facades." *Renewable and Sustainable Energy Reviews* 16 (2012): 2820–2832.

Saadatian, O., K. Sopian, C. H. Lim, N. Asim, and M. Y. Sulaiman. "Trombe Walls: A Review of Opportunities and Challenges in Research and Development." *Renewable and Sustainable Energy Reviews* 16 (2012): 6340–6351.

Said, I. "Carving Motifs in Timber Houses of Kelantan and Terengganu: Sustaining Malay Architecture Identity." In *Conference on Technology and Sustainability in the Built Environment*. Saudi Arabia: King Saud University, 2010.

Salomé, R. *Mediterraneidade – Interacções no Design de Produto: A Identidade Cultural como Referente para uma Actividade Projectual Sustentada*. PhD diss., University of Porto, Faculty of Engineering, 2012.

Schittich, C. *Solar Architecture*. Basel: Birkhäuser, 2003.

Smith, P. *Architecture in a Climate of Change*. Oxford: Architectural Press, 2001.

Wilson, E. *Diseños Islâmicos*. Mexico: Editiones G. Gili, 1998.

Zamora, B., and A. S. Kaiser. "Influence of the Variable Thermophysical Properties on the Turbulent Buoyancy-Driven Airflow inside Open Square Cavities." *Heat and Mass Transfer* 48 (2012): 35–53.

Index

Note: Figures are indicated by *italics*. Tables are indicated by **bold**.

R

S

T

U

V

W

For Product Safety Concerns and Information please contact our EU representative GPSR@taylorandfrancis.com
Taylor & Francis Verlag GmbH, Kaufingerstraße 24, 80331 München, Germany

www.ingramcontent.com/pod-product-compliance
Ingram Content Group UK Ltd.
Pitfield, Milton Keynes, MK11 3LW, UK
UKHW020202200926
475702UK00004B/570
9781041136170